Market Opportunities for African Agriculture

A General Equilibrium Examination of Demand-Side Constraints on Agricultural Growth in East and Southern Africa

Xinshen Diao
Paul Dorosh
Shaikh Mahfuzur Rahman

RESEARCH REPORT 154

INTERNATIONAL FOOD POLICY RESEARCH INSTITUTE
sustainable solutions for ending hunger and poverty
IFPRI®

International Food Policy Research Institute
2033 K Street, NW
Washington, D.C. 20006-1002, U.S.A.
Telephone +1-202-862-5600
www.ifpri.org

DOI: 10.2499/9780896291638RR154

Library of Congress Cataloging-in-Publication Data

Diao, Xinshen.
Market opportunities for African agriculture : a general equilibrium examination of demand-side constraints on agricultural growth in east and southern Africa / Xinshen Diao, Paul Dorosh, and Shaikh Mahfuzur Rahman.
p. cm. — (Research report ; 154)
Includes bibliographical references and index.
ISBN-13: 978-0-89629-163-8 (alk. paper)
ISBN-10: 0-89629-163-4 (alk. paper)
1. Agriculture—Africa. 2. International trade. I. Dorosh, Paul Anthony. II. Rahman, Shaikh Mahfuzur, 1968– III. International Food Policy Research Institute. IV. Title. V. Series: Research report (International Food Policy Research Institute) ; 154.
S472.A1D53 2007
382'.41096—dc22 2007016541

Contents

Tables

Figures

Foreword

In today's more integrated world economy, agricultural growth in Africa depends not only on raising productivity and increasing production, but on increasing the competitiveness of African agriculture in the global market and expanding its market opportunities within Africa. Unless demand increases, African agriculture cannot grow at a rate sufficient to reduce poverty and hunger to any substantial degree. The accelerated economic growth in Africa in recent years might offer increased opportunities for agriculture from domestic demand.

This research report focuses on demand-side constraints on African agricultural growth and their implications for three broad agricultural development strategies: promoting traditional exports, developing nontraditional exports, and increasing food-staple growth. Applying a general equilibrium framework to seven East and southern African countries, the study finds that an export-led agricultural growth strategy is unlikely to generate substantial overall income growth, even though demand need not constrain rapid growth in some nontraditional exports. Increasing production of staple foods seems to offer a promising avenue for agricultural growth, given that Africa's supply of many staple commodities does not meet its current demand. However, the report shows that increasing productivity in the grain sector without a substantial increase in market demand would cause domestic market prices to fall; as a result, farmers would lose much of the benefit of increased productivity. Thus, changes outside the farm sector itself—reduced marketing costs and more rapid growth in the nonfarm economy—are required to provide sufficient market demand to support rapid agricultural growth. In addition, to increase agricultural productivity, a second major engine of growth—increased productivity growth outside the agricultural sector—is a necessary condition for rapid economywide growth and poverty reduction.

Agricultural growth can play a major role in overall growth and poverty reduction in Africa. But development strategies need to take an economywide perspective to enable agriculture to play its crucial role in growth and poverty reduction in Africa.

Joachim von Braun
Director General, IFPRI

Acknowledgments

The analysis presented herein is part of a research project to support the U.S. Initiative to End Hunger in Africa (IEHA). The report draws upon and extends an earlier analysis published as DSDG Discussion Paper No 1. We thank Siet Meijer, Mark Rosegrant, Yukitsugu Yanoma, and Weibo Li for their contributions to the discussion paper. We also thank the many people who made important contributions to this research project, particularly Peter Hazell for providing overall guidance and many helpful suggestions during the entire research process. Eleni Gabre-Madhin, Michael Johnson, and Nicholas Minot also provided valuable input in brainstorming sessions during the early stages of the project. Steven Haggblade, Jock Anderson, and outside reviewers of an early version of the discussion paper manuscript, and the three anonymous reviewers of the present manuscript, provided helpful comments and suggestions. Thanks also go to Danielle Resnick, Sarah Cline, Christen Lungren, Liangzhi You, and Bingxin Yu, who assisted with data and other aspects of the research. Finally, we thank USAID for funding this research.

Summary

In today's more integrated world economy, agricultural growth in the countries of East and Southern Africa depends crucially on increasing their competitiveness in the world agricultural market and expanding their market opportunities. This report focuses on demand-side constraints on agricultural growth and their implications for three broad alternative agricultural development strategies: promoting traditional exports, developing nontraditional exports, and increasing food staple growth. We address three major questions. First, how constraining will demand be for future agricultural growth in East and Southern Africa and, in particular, is there sufficient demand to permit agriculture to grow at a rate that can significantly reduce poverty and hunger? Second, if technological change and increase in supply are achieved, which agricultural subsectors offer the greatest potential and can become the most powerful engine for raising real incomes and increasing food consumption? Finally, what are the implications of reductions in marketing costs and growth linkages with nonagricultural sectors for achieving increased market demand for agricultural products?

The report applies a general equilibrium framework for the analysis, focusing on seven East and Southern African countries—Madagascar, Malawi, Mozambique, Tanzania, Uganda, Zambia, and Zimbabwe—and finds that an export-led agricultural growth strategy is unlikely to generate substantial overall income growth in these countries. Despite significant market reform initiatives, future growth prospects for traditional agricultural exports in many countries in the two regions do not appear promising, even if lost market shares are recovered through improvements in productivity, product quality and variety, and marketing conditions. Simulation of a recovery of traditional exports to their historical high levels results in only an additional 0.08–0.15 percent annual growth in per capita real GDP for the seven countries studied. The limited impact of an increase in traditional exports is due primarily to the fact that they now account for only a small portion of the GDP and production of these commodities has relatively weak economic linkages with other sectors in the domestic economy. Moreover, world prices for these commodities are currently rather low and their recovery in real terms to the levels of the early 1970s is unlikely.

The report suggests that demand need not constrain rapid growth in nontraditional exports. Given the current minimal levels of production and their minute share of world trade, nontraditional agricultural exports offer perhaps the most promising opportunities for growth. To a large extent, nontraditional exports consist of numerous products that are targeted to niche markets and the model simulation analysis shows that even under optimistic assumptions (for example, 10–24 percent annual export growth in the sample countries), acceleration of growth in this sector alone will have only a very limited effect on real GDP growth.

Increasing production of staple foods would seem to offer a promising avenue for agricultural growth, given that Africa's own demand is already large ($18 billion/year for East African and $12 billion/year for Southern African developing countries) and that the supply of many staple commodities is not sufficient to meet the current demand. Moreover, the total demand for food in Africa is projected to increase in the next 10–20 years (for example, cereal demand will be 60 percent higher than the current level by 2015). However, an economywide model simulation analysis shows that increasing productivity in the grain sector alone would cause a shift in domestic terms of trade that would have a negative impact on agriculture as domestic market prices fall, which would cancel out most benefits accruing to the farmers from the improvement in productivity. Increased substitution of domestic agricultural products for imported commodities through government intervention in grain markets could mitigate these adverse trends in terms of trade for farmers. Simultaneous increase in livestock productivity might also increase farmers' incomes and spur demand for food and feed, which would make it more effective for boosting farm incomes than an increase in crop productivity alone.

Nonetheless, the model simulations suggest that changes outside the farm sector itself—reduced marketing costs and more rapid growth in the nonfarm economy—are required to provide sufficient market demand to support rapid agricultural growth. Model simulations suggest that a sharp reduction in marketing costs through investments in marketing infrastructure (for example, roads and bridges, ports, storage facilities, electricity) and development of market institutions (see, for example, Kherallah et al. 2002) combined with agricultural productivity growth would raise per capita GDP growth by approximately 2 percent per year in the sample countries. Simulation results also suggest that, combined with nonagricultural productivity growth, productivity growth in agriculture results in per capita annual agricultural real income growth of 3.0–4.4 percent.

Thus, agriculture, particularly productivity increases in food crops and livestock, can serve as one engine of growth, but without economywide growth, total demand for crops and livestock will be insufficient to prevent terms-of-trade declines. A second major engine of growth—increased productivity growth outside the agricultural sector—is a necessary condition for rapid economywide growth and poverty reduction.

CHAPTER 1

Introduction

The past 35 years have seen a dramatic shift in the international consensus on the importance of agriculture in economic development. From the 1960s through the 1980s, agriculture was generally recognized as having a central role. By the 1990s, however, interest in agriculture's role in development had waned, with adverse consequences for many African countries that continued to face recurrent food crises, economic stagnation, and persistent poverty (Hazell 2005). Today, agriculture is again a major item on the agenda for Sub-Saharan Africa, although there remains considerable doubt in the international development community as to whether it can successfully generate sufficient growth in Africa (see, for example, Maxwell and Slater 2003; Collier 2005; Ellis 2005).

Pessimism regarding the future of agriculture in Sub-Saharan Africa is the result of a change in view on appropriate policy measures, concerns about the robustness of domestic demand, and perceptions of the implications of a changing global environment. Government policy instruments that supported the Green Revolution in Asian countries by improving returns to agriculture, such as price supports, fertilizer and credit subsidies, and irrigation schemes, are less acceptable models for public sector intervention today (Ashley and Maxwell 2001). Domestic demand, dampened by low incomes and high market transaction costs, is often deemed inadequate to translate increased production of food staples into increases in rural incomes. Long-term global declines in agricultural commodity prices have also undermined the profitability of agriculture as a business. Moreover, African agriculture faces new challenges from globalization, resulting in increased competition from both other developing countries and wealthy countries with agricultural subsidies. Growing concentration and integration of international markets, as well as increasingly demanding standards in terms of quality and food safety, pose additional challenges (Dolan and Humphrey 2000; Reardon et al. 2003; Weatherspoon and Reardon 2003).

What can be done to enhance market opportunities so that agriculture can become a more powerful engine for growth for the continent? Which markets and which products offer the greatest potential for raising incomes and increasing food consumption in Africa? To answer these questions requires an analysis of demand constraints on agricultural growth. Much of the earlier work on agricultural growth focused on means of boosting supply of agricultural products, a crucial first condition for growth. African farmers clearly need more public and private investment to provide additional resources for research and extension, for promoting rapid adoption of new technologies, and for adapting modern technology to local agroecological and socioeconomic conditions. However, lack of market opportunities for increased production and poorly functioning markets can also constrain agricultural growth, so recent studies have focused on the demand side and the role of markets in African agricultural growth. Dorward, Kydd, and Poulton (2005), for example, have pointed out the importance of market coordination and argue that coordination failure, by increasing transaction costs and discouraging market participation, is one of the most important factors behind the poor performance of the African agricultural sector.

In this report, we focus on demand-side constraints and their implications for three broad alternative agricultural development strategies in East and Southern Africa: promoting traditional exports, nontraditional exports, and food staple growth. The purpose of this analysis, however, is not to derive country-specific policy recommendations, but rather to examine development strategy issues faced by many countries in these regions. Nonetheless, the analysis does include some findings concerning particular countries, subject to the limitations of the dataset used for the general equilibrium modeling.

Specifically, this report is designed to address three major questions faced by many countries in East and Southern Africa: (1) How constraining will demand be for future agricultural growth and is there sufficient demand to permit agriculture to grow at a rate that can significantly reduce poverty and hunger?[1] (2) If technological change and increases in supply are achieved, which agricultural subsectors have the best potential for raising real income and increasing food consumption? (3) What are the implications of reductions in marketing costs and growth linkages with nonagricultural sectors for achieving increasing market demand for agricultural products?

Through a combination of analyses of historical trends in agricultural exports, comparative advantages in intraregional trade, and various growth scenarios in a world general equilibrium model, we examine the impact of linkage between growth in agricultural demand and supply on overall economic growth and income, taking into account the internal demand constraints and diverse economic structures across a group of countries in East and Southern Africa. We find that traditional export crops face overall demand constraints that are reflected in long-term declines in terms of trade, although there is scope for some countries to recoup lost market shares. However, if the goal is to cut rural poverty by 50 percent by 2015 the potential income generated is small relative to the need.

Owing to their relatively small base in many countries, nontraditional exports may be subject to fewer demand constraints than traditional exports, but they have limited potential to raise incomes on the scale required to affect overall economic growth and reduce poverty. The best prospects for agriculture-led growth in many African countries remain in the food sector, where domestic demand still represents a large and growing market. Improvements in marketing efficiency and simultaneous growth in the livestock sector (and feed demand) can help spur demand further and prevent the price declines that discourage grain production. Achieving rapid gains in farm incomes, however, will require not only sustained increases in agricultural output, but investments in rural infrastructure to reduce marketing costs and enhance growth in nonagricultural sectors in order to produce the income necessary to spur demand.

In Chapter 2, we first examine trends in Sub-Saharan Africa's traditional and nontraditional agricultural exports and the structure of agricultural trade. Based on previous studies and recent experience of several African countries, we focus primarily on world export market opportunities and constraints for a group of selected traditional and nontraditional commodities. In Chapter 3, we look at regional and domestic demand for food products, analyze comparative advantages in regional trade, and evaluate the potential impact of regional and domestic markets on farm income and agricultural growth. Chapter 4 presents the general equilibrium model and discusses simulation results derived from the model. Concluding observations and policy implications are offered in Chapter 5.

---

[1]AfDB (2002) for example, estimates a 7 percent of annual GDP growth for achieving the MDGs of halving the poverty in Africa.

CHAPTER 2

Agricultural Export Markets: Trends, Constraints, and Opportunities

Agriculture is widely recognized as a key to growth and the reduction of poverty in Sub-Saharan Africa. Thus, the disappointing performance of the agricultural sector from the 1970s through the early 2000s in most of the countries in the region is seen as a major factor in increasing poverty. Although there were higher returns to some farmers in a few subsectors in some reforming countries, structural adjustment reforms introduced in the 1980s and 1990s failed to produce substantial agricultural growth, owing in part to inconsistent implementation (Dorosh, Sahn, and Younger 1996; Jayne et al. 1999; Kherallah et al. 2002). Declines in public investment in rural infrastructure, agricultural research, and agricultural extension also contributed to both stagnation in agricultural productivity and high market transaction costs.

Consistent with the disappointing performance of agricultural production, Sub-Saharan African agricultural exports, particularly the traditional ones, have fared badly in many countries. In this chapter, we first examine trends and prospects for these traditional exports, which still account for the major share of agricultural export earnings in most countries. We then discuss nontraditional exports, some of which have enjoyed extremely rapid growth, though from a limited base. Prospects for trade in staple crops and livestock, including intraregional trade, are discussed in Chapter 3.

Traditional Exports: Increased Competition with Slow Growth in Demand

Many Sub-Saharan African countries have a long history of exports of tropical tree crop products (cocoa from West Africa and coffee from several regions), as well as other traditional exports (cotton, tea, and tobacco). Aggregating from national level data reported by FAO (2006), we find that these traditional exports accounted for 60 percent of the total value of Sub-Saharan Africa's agricultural exports. The five main traditional export crops provided Africa with an average of $5.6 billion in annual revenues over the last 25 years,[1] a level about 40 percent higher than that in the 1970s. However, the world total export value of these five crops increased by 120 percent in the same period, so that Africa's market share fell from 22 percent in the 1970s to 14 percent in the 2000s. If African traditional exports had grown at the same rate as total world trade in these commodities, these traditional exports would have generated $8.8 billion in earnings each year to Sub-Saharan Africa as a whole, $3.2 billion more than actual earnings. A major reason for the slow growth in export earnings is the weak performance of coffee ex-

[1]All reported dollar figures are in US$.

Figure 2.1 Monthly average prices for coffee and cocoa, 1980–2005

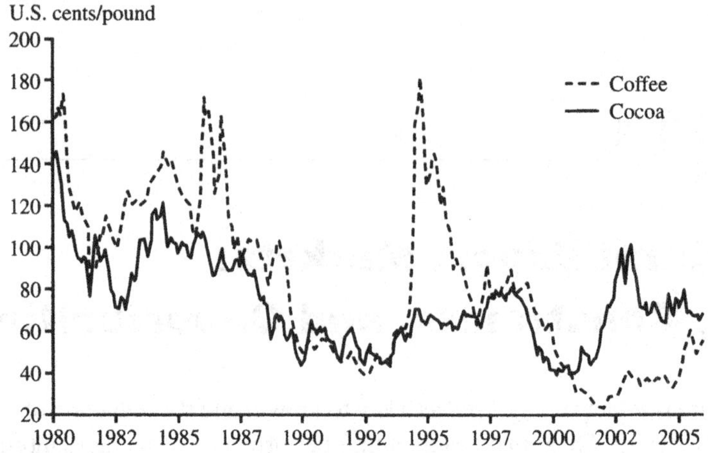

Source: ICO (2006) and ICCO (2006).

ports. If the figures for coffee are excluded, export revenue of the other four crops rose by 120 percent between the 1970s and the 2000s, compared with an increase of 190 percent in the world trade.

Weak global demand for traditional export commodities, particularly for coffee, is one factor behind the poor performance of African agricultural exports. Ng and Yeats (2002), for example, argue that global demand for the type of goods African countries produce has grown much more slowly than the growth in demand for other commodities. A study on economic development in Africa by the United Nations Conference on Trade and Development (UNCTAD 2001) argues that a secular decline in African terms of trade since the early 1980s is responsible for the marginalization of the region in world trade. According to the estimates of the UNCTAD study, the level of overall terms of trade for Africa at the end of the 1990s was 20 percent below that attained in the early 1970s. For the two largest African export commodities, coffee and cocoa, the lowest world market prices in 2000–2002 were only about 12 percent of the highest prices for coffee in the early 1990s and 27 percent of the highest prices for cocoa in the early 1980s (Figure 2.1). Declines in coffee prices alone represented a total loss of about $4.5 billion (ICO 2006). Given that the unit value of nonagricultural products (for example, manufactured goods) was rising during a similar period, the real prices for Africa's traditional export commodities were further lowered from their nominal values.[2] Moreover, the downward trend in relative prices of African export commodities was accompanied by a high degree of volatility. A study by Akiyama et al. (2003) shows that African exports experienced approximately twice the volatility in terms of trade as East Asia from the 1970s to the 1990s and roughly four times that experienced by the industrial countries during the same period.

Coffee

Coffee is produced in 40 Sub-Saharan African countries, 20 of which are in East and Southern Africa (Table A1).[3] The decline in green coffee prices is due mainly to an oversupply in the global coffee market resulting from increases in world coffee production, which are driven primarily by increases in yield in many countries(Figure 2.2), in addi-

[2]According to Baffes (2003), world cotton price in 2002 was 50 percent of its 1980s level in nominal terms and 40 percent in real terms.

[3]All table numbers including the letter A refer to the tables in Appendix A.

Figure 2.2 Coffee yield in Africa, Latin America, and Asia

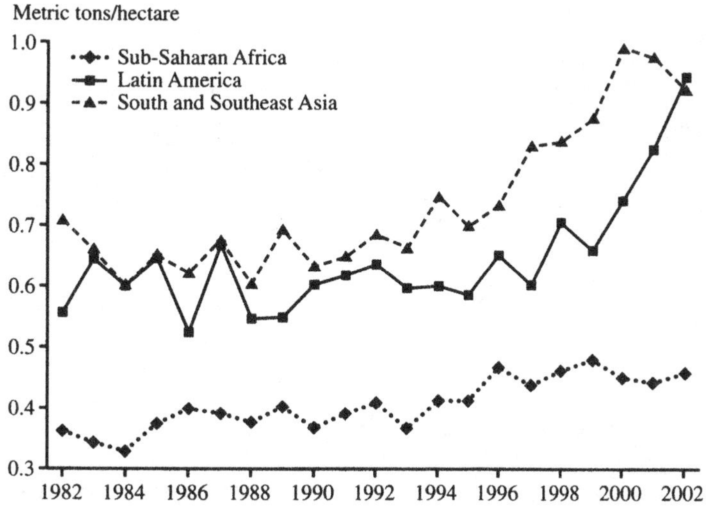

Source: FAOSTAT (2006).

tion to the area expansion in some countries. Owing to technological innovations that have resulted in high-yield varieties and better pest resistance, world average coffee yields grew by 5.5 percent a year between 1996 and 2003 (Table 2.1, Part A). For example, yield increased dramatically after the introduction of Caturra plants in 1996 in Colombia. In Brazil, coffee yield grew at 6.8 percent per year, and there was an increase of 6.6 percent per year for Latin America as a whole during this period (Table 2.1, Part C). The area expansion that occurred mainly in South and Southeast Asia (Table 2.1, Part D) has made a few countries in these regions major competitors in the world coffee market. Vietnam, for example, has increased its market share from almost zero in the 1970s to more than 10 percent today. Together with India and Indonesia, the share that Asian countries have gained in the world coffee market is almost equivalent to the share that African countries have lost (Table 2.1, Part E).

In contrast, green coffee yields either declined or stagnated in the same period (1996–2003) in 25 of the 28 Sub-Saharan African countries for which the yield data are available. For Sub-Saharan Africa as a whole the coffee yield declined by an average annual rate of 0.3 percent during the same period (Table 2.1, Part B). A cross-country yield comparison for different types of coffee shows even more dramatic changes.[4] Figure 2.3 compares the trends in robusta coffee yield in Côte d'Ivoire and Vietnam, and Figure 2.4 compares arabica coffee yields in Kenya, Tanzania, and Colombia. Only for the hard arabica group did yields in Ethiopia remain higher than in a major competing country (Brazil) until 2001, although Brazilian yields seem to have a much higher growth rate (Figure 2.5). These yield comparisons suggest that declines in coffee production efficiency is one major reason for Africa's loss of market share.

Moreover, in 24 Sub-Saharan African countries, harvested areas either declined or stagnated in this period.[5] In total, Africa lost one-third of its coffee production area (an annual average decline of 3.2 percent; Table

[4]International Coffee Organization has categorized coffee into four major groups: Colombian mild arabica, Brazilian natural or hard arabica, other mild arabica, and robusta. Three types of coffee, Colombian mild arabica, hard arabica, and robusta, are produced in Africa.

[5]Coffee production in Angola, one of the world's largest coffee producers during the 1960s and early 1970s, fell to almost zero in the 1990s owing to civil war (Table A1). An outbreak of the coffee wilt disease is an important reason for the fall in Uganda's coffee production (Baffes 2006).

Figure 2.3 Robusta coffee yield in Côte d'Ivoire and Vietnam

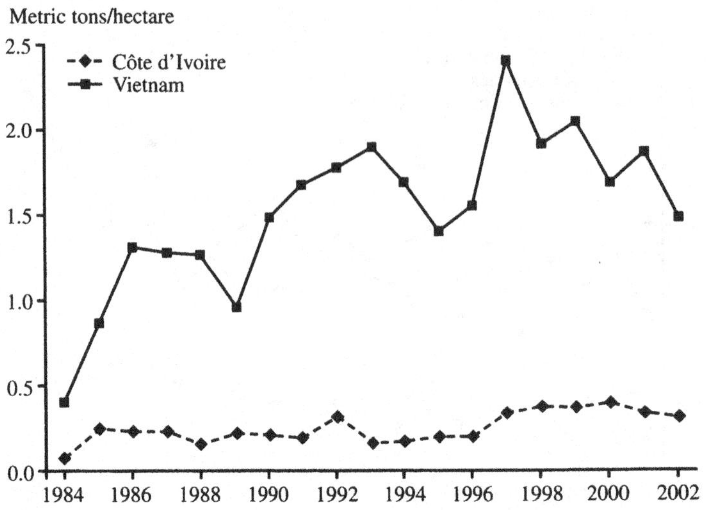

Figure 2.4 Colombian mild arabica coffee yield in Kenya, Tanzania, and Colombia

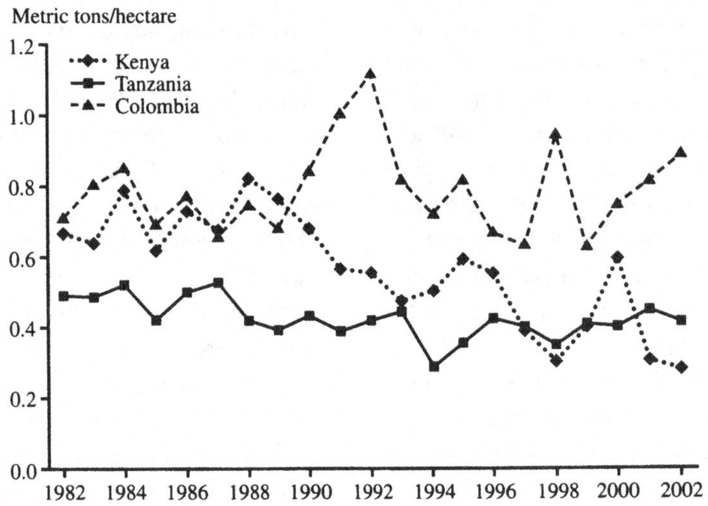

Figure 2.5 Hard arabica coffee yield in Ethiopia and Brazil

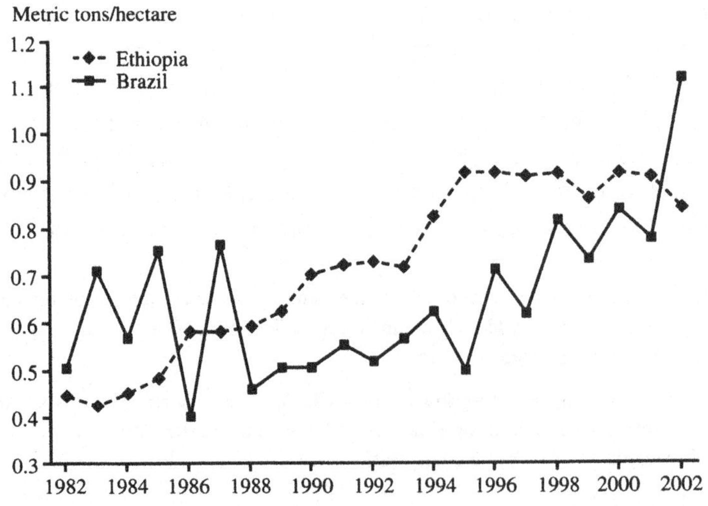

Table 2.1 Growth rate of coffee production and exports (annual average %)

	1970–79	1980–89	1990–95	1996–2003
A. World				
Production growth rate	1.0	1.5	–2.0	6.3
Yield growth rate	0.5	0.2	0.6	5.5
Harvest area growth rate	0.5	1.3	–2.7	0.7
Export volume growth rate	0.3	2.2	–2.0	2.5
Export value (current prices) growth rate	20.3	0.4	13.2	–14.0
Share of coffee in world agricultural trade	5.3	4.4	2.2	2.1
B. Sub-Saharan Africa				
Production growth rate	–2.0	0.7	–2.9	–3.4
Yield growth rate	–2.1	1.0	1.7	–0.3
Harvest area growth rate	0.1	–0.3	–4.5	–3.2
Export volume growth rate	–1.0	0.2	–5.7	–5.5
Export value (current prices) growth rate	19.2	–1.6	9.0	–19.8
Share of coffee in Sub-Saharan Africa agricultural trade value	24.8	25.9	14.2	11.2
C. Latin America				
Production growth rate	1.6	1.4	–3.9	7.6
Yield growth rate	1.4	–0.2	–0.8	6.6
Harvest area growth rate	0.2	1.6	–3.1	0.9
Export volume growth rate	–0.7	1.7	–3.4	2.9
Export value (current prices) growth rate	18.9	–0.1	11.0	–13.4
Share of coffee in Latin American agricultural trade value	22.8	20.6	13.3	11.1
D. South and Southeast Asia				
Production growth rate	5.7	3.7	5.7	10.1
Yield growth rate	1.3	–0.9	2.6	4.3
Harvest area growth rate	4.3	4.7	3.1	5.5
Export volume growth rate	8.9	6.2	3.1	7.1
Export value (current prices) growth rate	31.1	3.9	26.3	–13.4
Share of coffee in South and Southeast Asian agricultural trade value	3.9	5.2	3.6	3.3
E. Share in the world coffee market				
Sub-Saharan Africa (%)	30.6	23.8	18.8	15.8
Latin America (%)	57.7	58.8	58.0	53.9
South and Southeast Asia (%)	6.8	11.9	16.2	22.0

Source: FAOSTAT (2006), last accessed in March 2006.

2.1, Part B). With low productivity in coffee production Africa was less competitive in the world market. Among the 40 coffee-producing countries in Sub-Saharan Africa, 26 listed in Table A1 had annual coffee export revenue of more than $100,000 in the early 1970s. Today there are only four countries whose annual coffee exports are valued at more than $100,000. Among these four, only Ethiopia has a current annual coffee export revenue higher than it was in the 1970s, whereas revenues from coffee exports for the

Table 2.2 World shares and growth rates of roast coffee exports (annual average %)

	1970–79	1980–89	1990–95	1996–2003
World roast coffee exports				
As percent of green coffee exports	1.4	2.6	5.2	6.7
Growth rates	7.1	8.7	4.1	7.1
Shares in the world roast coffee market				
Sub-Saharan Africa	0.6	1.0	0.3	0.5
Asia	1.3	2.0	4.0	2.1
Europe	62.5	75.6	79.1	77.8
North and Central America	35.3	19.6	12.7	18.1
Growth rates of roast coffee exports				
Sub-Saharan Africa	10.0	36.7	0.4	–5.8
Asia	13.3	16.7	21.7	1.7
Europe	9.1	10.7	5.0	6.8
North and Central America	3.0	–2.4	3.5	9.4

Source: FAOSTAT (2006), last accessed in March 2006.

other three big coffee export countries, Côte d'Ivoire, Kenya, and Uganda, fell by 50–70 percent from the levels in the early 1970s.

According to FAO estimates, demand for coffee in developed countries (which consume about 70 percent of the world's coffee) is more or less constant with a low price elasticity (around –0.2); that is, demand will increase only marginally even if the price declines substantially. Without considerable growth in demand, even a modest growth in Africa's productivity (not to mention the continuing increase in coffee production in Latin America and Asia) may further depress world prices for green coffee.[6] There is, however, a long-term prospect for growth in demand in less mature markets of Asia, Eastern Europe, South America, North America, and in domestic markets of the producer countries themselves, where demand is highly income elastic. However, African countries have to compete with Asian and Latin American exporters for opportunities in these markets.

Although the global market for green coffee is characterized by sluggish demand, a rapidly expanding market for high-quality and differentiated gourmet coffees in recent years appears to offer a better prospect. "Specialty" coffee products differentiated by superior variety and higher quality are sold at a significant price premium on the international market; Colombia's Juan Valdez and 100% Colombian Coffee and Jamaica's Blue Mountain Coffee, for example, are the most expensive coffees in the world. The success of these brands is attributed to high quality and effective marketing strategies, but African countries have yet to develop such specialty products.[7]

There also remains a great potential for expanding the market for roast coffee. World export of roast coffee has been growing in terms of both volume and value. In the 1970s, the volume of world exports of roast coffee was equivalent to only 1.4 percent of that of green coffee. With an average annual

[6]According to DECPG (2006) coffee prices in the world market fell by 5.7 percent in June 2006 on expectations of a large increase in Brazil's crop for the 2006/07 season.

[7]Likewise, "Fair-Trade" coffee remains a niche market. Calo and Wise (2005) conclude that while fair trade coffee may have an important role to play in promoting sustainable livelihoods, neither this market nor national government programs alone can substitute for "concerted international efforts to address the price crisis." See also Bacon (2005).

Table 2.3 Growth rate of cocoa production and exports (annual average %)

Region	1980–89	1990–2003	1980–2003
Sub-Saharan Africa			
Production growth rate	5.0	3.7	4.2
Export volume growth rate	4.7	3.3	4.0
Export value growth rate	0.4	5.0	1.2
Share of cocoa in Sub-Saharan Africa's agricultural trade value	19.2	15.5	17.2
World			
Production growth rate	4.4	2.1	3.4
Export volume growth rate	6.1	2.4	3.7
Export value at current prices growth rate	1.3	4.2	0.9
Share of cocoa in world agricultural trade value	1.1	0.6	0.6
Share of Sub-Saharan Africa in world cocoa market	64.9	64.0	61.1

Source: FAOSTAT (2006), last accessed in March 2006.

growth rate of 8.2 percent, that ratio has now risen to 6.7 percent (Table 2.2). African countries started to increase roast coffee exports in the mid-1980s. However, owing to the extremely low level at which they began and a slowdown in growth in recent years, African countries currently account for less than 1 percent of world roast coffee exports (Table 2.2), equivalent to 1–2 percent of their green coffee exports.

Since world prices for roast coffee are about twice those for green coffee, African countries would be better off exploring market opportunities for roast coffee and increasing roast coffee exports. For example, if African countries could replace 15 percent of their current green coffee exports with roast coffee, assuming that the price ratio remains the same, it would be equivalent to an increase in coffee export revenue of 10 to 15 percent and total agricultural export value of 2 percent. Nevertheless, the potential depends crucially on how competitively African countries can roast their green coffee maintaining high quality compared with other countries that export roast coffee.

Cocoa

Cocoa is produced primarily in West Africa, but given its importance in Sub-Saharan Africa's total agricultural exports, we include it in this study and compare its trends with coffee. As opposed to the current situation with coffee, West African countries as a whole have been able to keep their share of the world cocoa market at about 60 percent, approximately the same level as their historical high in the 1970s. World cocoa production and exports have both grown at an average rate of 3.4 and 3.7 percent per year, respectively, in the last two decades, whereas West African cocoa production and exports have grown at a relatively faster pace, by 4.2 and 4.0 percent per year, respectively (Table 2.3). However, world cocoa prices declined significantly in the late 1990s (see Figure 2.1), resulting in a decline in West African cocoa export revenue during 1996–2003. With the new revelations concerning the health benefits of cocoa consumption and the recovery in world cocoa prices after 2000, the value of Africa's cocoa exports almost doubled between 2000 and 2003 (and fell slightly in 2004). A few big cocoa-exporting countries in West Africa, including Côte d'Ivoire, Ghana, and Nigeria, have benefited significantly from the recent cocoa boom. However, even in these countries (except for Côte d'Ivoire), the recent cocoa export revenues are just slightly higher than the recorded historical highs of the late 1970s.

Table 2.4 Growth rate of cotton, tea, and tobacco production and exports (annual average %)

	Sub-Saharan Africa				World			
	1970–79	1980–89	1990–95	1996–2003	1970–79	1980–89	1990–95	1996–2003
Cotton								
Production growth rate	–2.8	6.5	0.9	0.6	0.7	2.3	–0.5	0.0
Export volume growth rate	–4.2	7.4	1.3	3.8	0.5	2.4	3.6	1.7
Export value growth rate	6.6	4.1	–0.7	–5.8	11.5	0.3	2.6	-7.1
Share of cotton in agricultural trade value	8.9	8.4	11.4	9.1	4.1	2.9	2.2	1.5
Share of Sub-Saharan Africa in world market	12.9	12.1	12.7	14.3				
Tea								
Production growth rate	6.1	5.0	2.4	3.3	4.3	3.4	0.9	2.2
Export volume growth rate	6.1	4.5	3.2	–4.2	2.2	2.7	–1.5	2.2
Export value growth rate	16.6	2.9	2.4	–5.0	13.8	2.5	–4.2	–0.6
Share of tea in agricultural trade value	2.5	3.7	4.9	4.6	1.0	0.9	0.7	0.6
Share of Sub-Saharan Africa in world market	17.7	20.0	24.4	21.0				
Tobacco								
Production growth rate	4.9	2.8	4.5	–3.4	2.7	1.9	–3.1	–4.5
Export volume growth rate	4.2	0.5	10.6	–1.3	3.8	1.0	4.9	0.0
Export value growth rate	12.2	2.4	4.1	–3.1	15.3	5.3	3.5	–4.5
Share of tobacco in agricultural trade value	3.4	5.0	9.4	8.6	3.7	4.1	5.7	4.7
Share of Sub-Saharan Africa in world market	7.9	9.1	11.0	9.1				

Source: FAOSTAT (2006), last accessed in March 2006.

Cotton, Tea, and Tobacco

In contrast with a declining trend in African coffee production and exports and a stagnating growth pattern in cocoa export revenue, African farmers as a whole have performed quite well in the other three traditional commodities in the last two decades. During this period, the annual growth rates of production, export volume, and export revenue of African cotton, tea, and tobacco remained above the world average (Table 2.4). Although the decline in the world price for cotton in the late 1990s and first 2 years of the 2000s caused African cotton export revenues to fall, exports grew at 3.8 percent a year in volume and 2.7 percent a year in absolute value between 1980 and 2003. In contrast, the world cotton trade grew at 1.5 and 0.3 percent a year in volume and value, respectively, during the same period. African tea production and exports also performed better than the world average, resulting in a growing share in the world tea export market (Table 2.4). Growth in African tobacco exports was also above the world average level, but stagnant world demand for tobacco started to constrain the growth of African tobacco exports.

Although Sub-Saharan Africa as a whole has generally been able to compete with other suppliers in cotton, tea, and tobacco export markets, subsidies in other countries, especially in several wealthy ones, have placed many African countries at a competitive disadvantage. According to the estimate of Townsend and Guitchounts (1994), more than two-thirds of the cotton produced in the world was grown in countries where there was some type of government intervention. The International Cotton Advisory Committee (ICAC 2002) found that eight countries provided direct support for cotton production and that the level of direct production assistance during the five seasons between 1997/98 and 2001/02 ranged between \$3.8 and \$5.3 billion. For 2001/02 alone, direct assistance to cotton producers in the United States reached \$2.3 billion, and \$0.8 billion went toward subsidies in Greece and Spain, the two major cotton-producing countries in the European Union (Baffes 2003). Market opportunities for African cotton and other traditional export crops would depend on the removal of such subsidies in other cotton-producing countries, especially the United States. Although Africa has the potential to increase exports in a less price-distorted world cotton market, competition from India, Brazil, and China may still constrain its market opportunities.

Reforms in the Traditional Agricultural Export Sector

Historically, production and marketing of traditional agricultural export commodities in most African countries has been government controlled, with the support of quota or buffer stock programs organized through international commodity agreements. The collapse of international prices for coffee, cocoa, and tea in the late 1970s and early 1980s made the quota and buffer stock programs unsustainable and signaled the end of those agreements. Moreover, the emphasis on market liberalization under the World Bank's structural adjustment programs led to the elimination of price controls and marketing boards for many traditional export commodities. By the end of the 1980s and into the early 1990s, the state monopolies in different African countries that implicitly taxed coffee and cocoa producers gave way to liberalization. In the case of cotton, policy approaches differed significantly between West and East Africa, owing to the very different economic and political conditions in the two regions.

In the prereform period, agricultural prices were typically severely depressed in many African countries, with negative protection often compounding the effect of overvalued exchange rates. Thus liberalization of agricultural prices was a common feature of reform programs. In the wake of the reforms, with the assumption that well-functioning agricultural markets existed or would spring up quickly and automatically, improving overall economic efficiency was expected to increase producer prices relative to border and international prices as a result of increased competition among traders as well as lower taxes. Since a competitive environment can stimulate investment and improve management, it was also predicted that transformation costs (for example, processing, storage, and transportation) in agricultural commodity marketing would be lowered and margins of profitability in the various segments of the marketing chain (from the world market to the border, and from the border to the farmgate) would decline.

Measures to improve markets and more generally to increase the elasticity of supply were totally neglected in the structural adjustment programs (Noman and Stiglitz 2006). Moreover, as changes in world prices now impact on the domestic commodity markets more directly and quickly, there is the question of whether market reform leads to greater variability in export prices. Several studies have attempted to test these predictions. Mundlak and Larson (1992) estimated price transmission elasticities of world to domestic markets for 58 countries and found that they were relatively low in the distorted domestic markets. Cardenas (1994), reaching a similar conclusion, noted that the fluctuations in world prices were found to fully transmit to producers where government intervention in pricing was limited (for example, in Costa Rica and Kenya). On the other hand, the findings by Quiroz and Soto (1995) differ substantially from these studies, as they found that the international price signals were transmitted poorly to domestic markets or not transmitted at all. Using a similar method, Baffes and Gardner (2003) find that changes in world prices account for only a small share of the variation in domestic prices. They argue that it was persistent political impulse that insulated domestic markets from world commodity markets. Krivonos (2004) reports that in most countries the long-term share of producer prices in world market prices increased substantially after the reforms, especially in countries where the liberalization was complete (for example, in Uganda, Brazil, and India).

Unlike numerous studies concerning changes in spatial prices after the reforms, studies relating changes in price volatility following reforms are rare. Akiyama et al. (2003) calculated variability in producer prices for cocoa, coffee, cotton, and sugar in 35 African countries for 1986–90 and 1991–95. Using FAO data, they found no evidence that price volatility increased between the two periods. Analyzing aggregate data and distinguishing between volatility and uncertainty, Dehn (2000) found evidence that commodity price uncertainty for African countries increased after 1973. However, he did not find any evidence that price uncertainty increased during 1986–97 compared to 1973–85. In a one country–one commodity study, Karanja (2002) found that price volatility in domestic coffee prices increased in Kenya following the reforms.

Apart from the resulting changes in the level of absolute producer price and price volatility, there is a need to evaluate the impact of agricultural commodity market reforms on the distribution of income and total surplus and on the structure of the market. It is observed that producers' share in the world price increased in almost all of the African countries that undertook significant reform initiatives. For example, prices received by Ugandan coffee producers moved much closer to world prices after the reforms (Figure 2.6), and a similar situation is observed in the coffee sectors in Ethiopia, Kenya, and Tanzania. In contrast, the margin between the world price and the price received by the robusta coffee producers in Côte d'Ivoire remained high, as the Ivorian coffee sector was not liberalized until 1998–99 (Figure 2.7).

The differences in marketing costs and taxes are also reflected in the prices received by producers. When the farmgate

Figure 2.6 World price and price received by Ugandan producers for arabica coffee

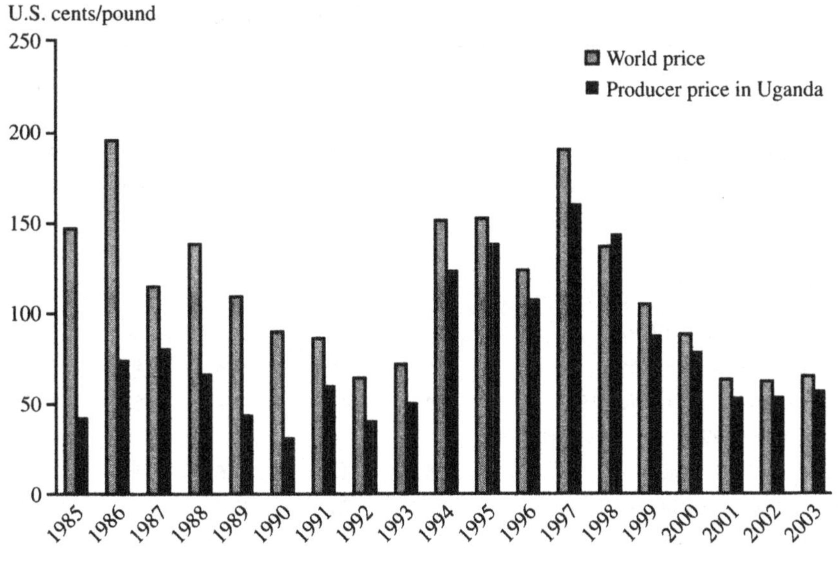

Source: Authors' calculation using ICO (2006) data.

Figure 2.7 World price and price received by Côte d'Ivoire producers for robusta coffee

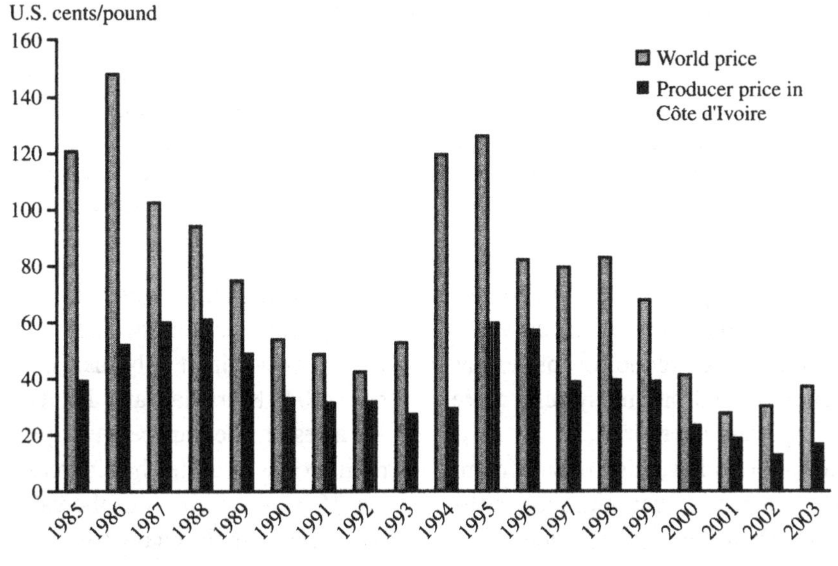

Source: Authors' calculation using ICO (2006) data.

prices were set by the marketing boards or price stabilization funds, farmers had no alternative outlet for their products, so the high marketing costs were easily passed on to the producers (Varangis and Schreiber 2001). Table 2.5 shows that African cocoa producers in the countries where cocoa marketing and pricing systems were controlled by state parastatals received much lower prices than the growers in countries with effective competition among buyers. In the wake of market reforms in 1998–99, producer prices in Côte d'Ivoire increased significantly. In Nigeria and Togo, producer prices increased from 20 percent and 60 percent of the FOB prices prior to the re-

Table 2.5 Cocoa producer price and producers' share of world price

	1991	1994	1996	1998	2000	2001	2002
World price (U.S. cents/lb)	54.2	63.3	66.0	76.0	40.3	49.4	80.7
Brazil (reform year: 1990)							
Producer price (U.S. cents/lb)		34.1	46.8	60.1	32.2	45.1	29.6
Producer's share of world price (%)		53.8	70.9	79.1	80.0	91.3	36.7
Cameroon (reform year: 1995)							
Producer price (U.S. cents/lb)	35.1	19.1	27.8	52.6	29.0	18.3	50.9
Producer's share of world price (%)	64.7	30.2	42.1	69.1	72.1	37.1	63.1
Côte d'Ivoire (reform year: 1999)							
Producer price (U.S. cents/lb)	35.1	26.8	27.8	34.8	23.5	42.2	47.9
Producer's share of world price (%)	64.7	42.3	42.1	45.7	58.3	85.4	59.4
Ghana (reform year: 1993)							
Producer price (U.S. cents/lb)	26.1	13.3	24.6	35.2	14.5	21.6	33.4
Producer's share of world price (%)	48.1	21.0	37.3	46.3	36.0	43.7	41.4
Indonesia (reform year: 1997)							
Producer price (U.S. cents/lb)	33.8	29.0	39.5	35.7	39.8	37.2	46.3
Producer's share of world price (%)	62.3	45.9	59.8	46.9	98.8	75.3	57.4
Nigeria[a] (reform year: 1987)							
Producer price (U.S. cents/lb)	46.8	126.4	166.6	165.3	34.8	36.7	46.6
Producer's share of world price (%)	86.4	199.6	252.3	217.5	86.4	74.4	57.8
Togo (reform year: 1996)							
Producer price (U.S. cents/lb)	43.9	25.5	39.1	50.1	24.8	47.6	59.2
Producer's share of world price (%)	80.9	40.3	59.2	65.9	61.6	96.4	73.4

Source: Compiled from ICCO (2006) and FAOSTAT (2006), accessed July and March 2006, respectively.
[a]Unusual increase in producer prices in Nigeria during 1994–98 was a result of exchange control.

forms to about 80 percent after (Varangis and Schreiber 2001). However, producers in Ghana continued to get lower prices, as cocoa marketing and pricing systems were still not completely liberalized.

Market reforms have also been accompanied by an increasing concentration of the downstream chain in certain commodity markets such as coffee and cocoa.[8] When millions of small producers in many low-income countries face a very few multinational companies that control the processing, marketing, and retailing, the belief is that asymmetric market powers contribute to the dramatic decline in international prices for green coffee, since the retail coffee prices declined only marginally (Krivonos 2004; Muradian and Pelupessy 2005). As a result, the gap between producer and retail prices of coffee has increased over the last 15 years and most of the value added in the coffee supply chain has been captured by the multinational companies (Talbot 1997). Analyzing the recent trends in the global coffee value chain, Ponte (2001) and Fitter and Kaplinsky (2001) conclude that even for the higher-quality and new variety coffees, the gains from product differentia-

[8]Five multinational companies became dominant in retail coffee sales in major consuming countries after the mid-1980s and seven large trading companies gained control over a major portion of the green coffee imports (Talbot, 1995–96). In 1998, the top six importers (Neumann, Volcafe, Cargill, Esteve, Aron, and Man) accounted for more than 50 percent of total global trade of green coffee, while the top two roasting and instant coffee manufacturing companies (Nestlé and Philip Morris) controlled 50 percent of the market (Ponte 2001).

tion do not necessarily accrue to the producers, as the downstream of the supply chain is becoming increasingly concentrated.

The market reforms in many African countries have also left small farmers without access to credit, price and production risk management tools, new technology, and some inputs such as insecticides and fungicides. This is an important factor for African small farmers unable to compete with large producers in Latin America or producers in Asia. Although unconventional production practices such as contract farming have been on the increase in the African traditional agricultural export sector, smallholders are often left behind as exporters prefer contracting with a small number of larger producers to minimize transaction costs. An empirical study on the Tanzanian coffee subsector, for example, shows that small producers became more vulnerable to financial constraints after the withdrawal of government support programs (Temu and Winter-Nelson 2002). Moreover, small growers' access to information and technical assistance worsened after the reforms because of the decline in government funds directed toward agricultural extension services (FAO 2000).

In summary, agricultural market reforms in many African countries seem to have had four fundamental impacts on the traditional agricultural export sector. First, as private trading is allowed in the domestic market and export taxes are substantially lowered, prices received by producers in the exporting countries have increased, at least in terms of the producers' share in export prices, and marketing margins have declined. Second, as price stabilization is abandoned, fluctuations in world prices have been transmitted more directly to domestic markets, exposing producers to a potentially higher degree of price volatility. Third, a major portion of the income and surplus generated along the traditional commodity chain was transferred from the producing countries to the multinational companies, as the market power shifted from the former to the latter after the collapse of international commodity agreements and increased concentration in the downstream part of the marketing chain. Finally, in the wake of the withdrawal of government support programs, African smallholders are facing the risk of being driven out of the market.

African countries' postreform experience suggests that simply opening up of the markets or elimination of government controls is not enough to bring about the desired growth of the traditional primary commodity sector. In order to be able to compete in a free-market environment Africa has to develop domestic market institutions and local infrastructure that facilitate the efforts of small producers. Although there remains some scope for increasing exports of traditional agricultural commodities and recapturing part of the lost world market share, the potential is limited by an almost stagnant global demand. Under these circumstances, many policymakers and researchers emphasize the expansion of nontraditional agriculture as a viable means of agricultural diversification, rural employment generation, and foreign exchange earnings in Africa. A few East African countries (for example, Kenya, Tanzania, Uganda) have already started to diversify their agricultural exports, moving away from the traditional commodities, but the actual growth potential of Africa's nontraditional agricultural exports has to be looked at very closely.

Nontraditional Exports

Nontraditional agricultural exports—commodities that are not part of conventional local production and are growing rapidly owing to newly discovered markets (Little and Dolan 1998; Ng and Yeats 2002; Singh 2002)—include a long list of products that vary across African countries and markets.[9]

[9]Some traditional export crops are considered as nontraditional in certain countries, such as cotton in Zambia.

In this study, we focus on two broad groups: horticultural and fish and fishery products.[10] Horticultural exports include more than 100 varieties of fresh and processed fruits, vegetables, cut flowers, and foliage, and exports of fish and fishery products comprise more than 20 species of fresh and processed freshwater and marine fish, cephalopods, crustaceans, and mollusks.

The destinations of the majority of African nontraditional commodities are the markets in developed countries, especially in Europe, where there is an increasing tendency for consumers to change their diets in favor of low-fat, low-cholesterol foods. Increased demand for healthy foods—fish, fruits, and vegetables—has led to a growth in imports from developing countries, where conditions for increasing production are likely to be more favorable. African countries not only have a comparative advantage in producing many nontraditional commodities, but they also have preferential access to the markets in developed countries, especially in Europe, under the trade agreements of the Lomé Convention of the World Trade Organization (WTO). However, future prospects of Africa's nontraditional agricultural export sector also crucially depend on whether African countries can cope with various nontariff barriers, including product quality and health and sanitary standards, and their durability in the buyer-driven supply chain.

Fish and Fishery Products

Fish and fishery products are still Sub-Saharan Africa's most important nontraditional agricultural exports. This subsector accounts for more than 15 percent of Africa's agricultural export value, increasing from 2 percent in the 1970s and 5 percent in the 1980s. With per capita consumption of 24 kilograms a year, the European Union is the largest market for fish in the world. Owing to a steady decline in its production of captured fish products since the late 1980s, the European Union has become increasingly dependent on imports, of which 50 percent are supplied by developing countries.

Although some 75 to 85 percent of African fish exports go to EU markets, such exports account for less than 8 percent of the total EU fish imports. The long-term trend suggests that consumption of fishery products in the European Union will continue to increase as will its imports, so there is tremendous potential for African countries to increase fish exports with the arrangement of duty-free access to the EU markets. However, the expansion of African fish exports is constrained by product species and nontariff barriers. For example, Sub-Saharan African countries produce and export mainly warm water fish species, which account for more than 70 percent of African fish exports, whereas EU fish imports are made up of primarily cold-water species. Crustacean products such as shrimp and prawns are the leading import group into the European Union, whereas their share in African fish exports is only 5 percent (European Union 2004).

Owing to the growing demand in the EU markets, there is opportunity for African countries to increase their exports of frozen, prepared, and preserved fish and processed fishery products, but nontariff barriers and changing supply channels are imposing constraints. Henson, Brouder, and Mitullah (2000) and Oyejide, Ogunkola, and Bankole (2000), for example, show that food safety requirements can easily act as barriers to fish trade. Exporting countries must qualify as "harmonized" or "provisionally harmonized" countries in terms of sanitary measures in order to gain access to EU markets. Import bans owing to food safety concerns are common in the EU countries in the fish

[10] We include fish as a nontraditional export category though it has been traditionally exported by some African countries.

trade. For example, the European Union imposed three successive bans on fish imports from East African countries between 1997 and 2000, causing disastrous outcomes in their fish industry, which is heavily dependent on the European market (Henson, Brouder, and Mitullah 2000).

African fishery products also have to traverse a complicated supply channel—African producers/processors, exporters, and European importers, end-product manufacturers, and retailers. Typically, the European importers are in a better position to govern the supply chain than the African processors or exporters. Moreover, multiple retailers have increased their market power in recent years and have become interested in acquiring certain products directly from overseas suppliers. Although changes in the structure of the supply channel may offer opportunities for African fishery exporters to work more directly with other parties in the chain, the requirement of a smoothly operating cold chain with adequate logistics to ensure consistent delivery of high-quality products is creating new challenges for African exporters.

There is also tremendous potential for increasing intraregional trade in fish and fishery products, which currently accounts for less than 10 percent of African fish exports. The region as a whole imports large quantities of fish and fishery products, equivalent to 50 percent of African fish exports, and with the growth of regional income this figure is expected to increase. African countries should be able to take advantage of this intraregion export opportunity.

Horticultural Products

Developed countries have seen a substantial increase in consumer demand for horticultural products during the last decade, and high costs of labor and heating for greenhouses have caused them to look for outside sources, particularly for out-of-season products. Tropical climates, counterseasonal patterns of production, and relative geographical proximity to Europe all give African countries a comparative advantage in horticultural export (Islam 1990; Barrett et al. 1997). Several other factors have also contributed to the growth of the African horticultural export market. Trade agreements such as the Lomé Convention give preferential treatment to African exports in the EU market (Singh 2002). Many African governments have adopted export promotion policies by reforming their domestic markets and trade policies and improving the environment for private and international investment. African countries also get technological and logistic support from their international counterparts for the production and marketing of horticultural products. Several African countries have also developed regional economic cooperation to coordinate their business activities and share technical know-how, market information, and human capital to increase their competitiveness. As a result of these various initiatives, several of them have seen very rapid growth in horticultural exports during the last decade.

Fruits and Vegetables. Fruits and vegetables accounted for 8 percent of African agricultural exports in 1996–2002 and ranked second in African nontraditional exports after fish and fishery products (Table 2.6). Apart from South Africa, Côte d'Ivoire, Kenya, and Tanzania are the major African exporters and together accounted for more than half of Sub-Saharan African total fruit and vegetable exports in the last two decades. Benin, Cameroon, Ghana, and Guinea-Bissau in West Africa, Uganda in East Africa, and Zambia in Southern Africa have also emerged as exporters of fruits and vegetables in recent years and together they currently account for another 20 percent of Sub-Saharan Africa's fruit and vegetable exports.

There is scope for many African countries to increase their share in the world's fruit and vegetable export market, as the 8 percent share of Sub-Saharan Africa's total

Table 2.6 Growth in fruit and vegetable production and exports (annual average %)

	1970–79	1980–89	1990–95	1996–2002
Sub-Saharan Africa				
Production growth rate	1.6	2.5	2.1	1.5
Export volume growth rate	0.9	0.4	4.8	4.8
Export value growth rate	7.6	1.0	7.9	1.8
Share of fruits and vegetables in Africa agricultural export value	4.0	4.3	6.1	7.6
World				
Production growth rate	2.6	2.9	3.9	4.3
Export volume growth rate	4.0	3.1	5.6	3.3
Export value growth rate	15.6	5.6	6.0	–0.9
Share of fruits and vegetables in Sub-Saharan Africa agricultural export value	11.5	13.0	16.0	16.1
Share of Sub-Saharan Africa in world fruit and vegetable export value	2.0	1.2	0.9	1.2

Source: FAOSTAT (2006), last accessed in July 2006.

agricultural exports is much lower than the 16 percent share of the fruits and vegetables in world agricultural trade. If South Africa is excluded, Sub-Saharan Africa as a whole accounts for only 1 percent of world total fruit and vegetable exports, which is close to the market share for South Africa alone. The Kenyan experience suggests that African countries have the potential to substantially increase their fruit and vegetable exports. Kenya's earnings from fruit and vegetable exports have grown steadily from $13 million in 1970 to $68 million in 1980 and to over $200 million in 2001. Significant increases in yields owing to improved technology and inputs, market innovations by private entrepreneurs, foreign investment, and government assistance all played vital roles in this growth.

Encouraged by the remarkable success of countries such as Kenya and Côte d'Ivoire, other African countries with favorable climates have also launched programs for horticultural export crop development. Ghana, Tanzania, Zambia, and Zimbabwe have entered aggressively into the European fresh produce market within the last few years with reasonable success. However, the entry of more competitors into the market opens up the possibility of lower average prices for some products and a decline in sales revenues, although estimated elasticity of export revenues with respect to export volume for some fruit and vegetable commodities is much higher than that for the traditional agricultural exports (Hallam et al. 2004).

Cut Flowers and Foliage. The world trade in cut flowers is characterized by a high degree of concentration by products, destinations, and sources. Roses (47 percent) and carnations (20 percent) are the main products traded. Germany is the principal importer, followed by the United States, the Netherlands, and Japan. The Netherlands leads the world in importing cut flowers for the purpose of later reexporting them (70 percent of the country's auctioned cut flower imports are reexported). African countries are the principal suppliers outside of Europe to European markets, but intra-EU trade is still a dominant factor in the EU's cut flower market.

Cut flower exports from Africa to EU markets increased dramatically during the last decade. The value of cut flower and foliage exports from seven African countries (Côte d'Ivoire, Kenya, Mauritius, Tanzania,

Uganda, Zambia, and Zimbabwe) increased from $104 million in 1994 to $355 million in 2003, accounting for almost 50 percent of extra-EU imports. At present, Kenya is the leading exporter of cut flowers and foliage to Europe (31 percent of extra-EU imports in 2003), and has managed to keep itself up to date with changing consumer preferences in terms of flower varieties, assortments, and colors. However, as the industry is highly dependent on a few varieties (roses, statice, and alstroemeria) for its profitability, increased competitive pressures from other African countries have weakened Kenya's position in the European market for certain products such as standard carnations, once the dominant focus of its industry (Thoen et al. 2001).

The Kenyan experience well illustrates the nature of the cut flower industry and the ever-changing challenges to maintaining international competitiveness in the market. First and most importantly, the industry is dynamic and is unlikely to be influenced by public initiatives and government interventions. The primary role of government should be facilitative.[11] Second, the development of a competitive and sustainable cut flower industry requires coordination in several areas by the principal stakeholders: floricultural and supply chain training, development of environmental standards, standardization of packaging materials, cold-chain facility investment, collection and dissemination of market information, and fostering of floricultural research and advisory services. Third, there is a need for further clarification and harmonization of environmental standards and regulations and effective communication of guidelines to growers and exporters. A joint public–private initiative for technical support to help comply with such guidelines and regulations may be necessary.

Smallholders in Horticultural Production. Export horticulture is regarded as a vehicle for poverty reduction through the generation of income among smallholders, rural laborers on large farms, and unskilled and semi-skilled processing factory workers. Von Braun (1994) reports that production of export vegetables has created new employment opportunities, reduced the need to rely on uncertain off-farm employment, and increased household income of the smallest Guatemalan farmers. A gross margin analysis by Minot and Ngigi (2002) suggests that horticultural production can produce substantially higher returns per hectare than staple food crop production. Since about 60 percent of the fruits and vegetables exported by Côte d'Ivoire and Kenya are produced by smallholders, they are the principal beneficiaries of the resulting direct benefits. Moreover, the indirect benefits to the economy associated with horticultural exports—an increase in employment opportunities, institutional development, marketing efficiency in the domestic supply chain, and the multiplier effect of export income flow in the rural economy—are likely to be even greater than its direct effects. McCulloch and Ota (2002) find evidence that households involved in export horticulture are better off than nonhorticultural households, particularly in rural areas, owing to the employment opportunities generated on farms owned by the major exporters and the access to credit and extension services that they provide. Nonetheless, the production and marketing of horticultural commodities often entail far more risk for small farmers, as compared with the production and marketing of grain.

The studies referred to earlier indicate gains for smallholders' involvement in horticultural exports, but the number of small farms producing horticultural crops has been declining steadily. For example, in the early 1990s the majority of fruits and vegetables for export were grown by smallholders in Kenya (Harris 1992, cited in Hum-

[11]Private sector initiative has been predominantly responsible for the emergence and growth of the Kenyan cut flower industry. The role of the Kenyan government has wavered between facilitation and constraint.

phrey, McCulloch, and Ota 2004), but by the late 1990s that number had declined to 18 percent (Singh 2002). Such a trend can be partially explained by the changing structure of horticultural supply chains in which exporters that are competing for a relatively small number of European supermarkets prefer to deal with a few large commercial farms, buying in large quantities through long-term contracts. These integrated supply chains are better able to coordinate the air shipping, rapid transport, and cold chains needed to deliver highly perishable horticultural products to the markets.

Although the share of smallholders in the newly developed supply chain seems to have declined, small growers compete favorably against large commercial firms in organic crop production and in the production of the more labor-intensive crops that require the stacking and picking of individual pods (Collins 1995). At present, there is a significant unmet demand for organic produce in European markets that provides entry opportunities for smaller African exporters (Dolan, Humphrey, and Harris-Pascal 1999). However, for smallholder operations to be successful it is essential that there be a sufficient number of cooperating growers in close proximity, located in areas with good transportation facilities. Moreover, to include more smallholders in the export-oriented crop production requires that small farmers be provided with full information about the market, be given access to credit, and be educated regarding suitable production and quality-control techniques.

Recently, the importance of horticultural production in peri-urban areas has drawn considerable attention among researchers and policymakers. Owing to short crop cycles, high labor input, and the small land area required for effective cultivation, peri-urban production of fruits and vegetables is particularly suitable for small-scale farming. Moreover, relative proximity to cities gives peri-urban horticultural settlements marketing advantages in terms of transportation and transaction costs. However, the possibility of involving a large number of smallholders in such activity is constrained by changing land-use patterns that come in the wake of urban expansion and the overall sanitary standards commonly found in the peri-urban areas (Binns and Fereday 1996).

Summary

The share of exports in the world's agricultural GDP has been increasing over time, but the value of African agricultural exports shows a declining trend, owing primarily to a sharp reduction in revenues from the region's traditional exports, which account for about 60 percent of Africa's total agricultural exports. Apart from facing an overall decreasing trend in world prices for these traditional commodities, many African countries have also suffered from declining terms of trade and substantial price instability over the past two decades. In addition to the limited growth in global demand for these traditional agricultural commodities and increasing competition from other parts of the world, African countries are also constrained by low productivity, problems with quality control, and asymmetric agricultural and trade policies of wealthy nations (especially production subsidies), which are the major reasons for Africa losing ground in the world marketplace. Despite African countries' significant market reform initiatives, future growth prospects for the region's traditional agricultural exports do not appear promising unless they can better their competitive position by improving productivity, product quality and variety, and marketing conditions.

Tropical climate, counterseasonal patterns of production, relative geographical proximity to European markets, and preferential access to the markets of developed countries give African countries a comparative advantage in nontraditional agricultural exports, such as fishery products, fruits, vegetables, and cut flowers. For a long list of such commodities, demand constraints are not binding in the short to medium term, and there are more promising opportunities for growth. However, success in capturing the market depends crucially on whether Africa can survive in the altered supply chain that requires quality and consistency. A few African countries have successfully increased their nontraditional exports, but the magnitude of these export earnings is still small for Africa as a whole.

African countries obviously need more diversified agricultural exports in order to reduce income risks from falling and unstable prices for traditional products. However, there remains the question as to whether nontraditional agricultural exports have significant potential as an engine of growth. The question arises not only because the volume of nontraditional exports is small relative to Africa's total trade and agricultural income, but also because of the relatively weak linkages with domestic and regional demand and, hence, with the activities and income growth of the greater part of the smallholders.

CHAPTER 3

Prospects of Regional and Domestic Markets

Unlike traditional and nontraditional agricultural exports, Sub-Saharan Africa's domestic demand and production of food crops and livestock products far exceeds internationally traded quantities. An average African household spends 50–70 percent of its income on food: the total value of food consumed is approximately $20 billion per year in East Africa and $12 billion in Southern Africa (excluding South Africa). Although only part of the food produced in the region is marketed, with the rest consumed on the farm, it still represents a large and growing market that should offer growth opportunities for African agriculture and farm income. In this chapter we assess the demand for food products in East and Southern Africa and explore regional market opportunities by examining whether there is potential to increase mutually beneficial agricultural trade across different countries and regions in Africa.

Food Demand in East and Southern Africa

Agroecological conditions vary tremendously across Africa, but most of the countries have large crop and livestock sectors. In East and Southern Africa, cereals (mainly maize and sorghum in most countries) typically account for 40–50 percent of the area cultivated, whereas among West African countries root and tuber crops are as important as cereals.[1] Given that trading volume of root and tuber crops is very small in East and Southern Africa, we focus on cereal crops to discuss regional and domestic market opportunities.

Although economists and policymakers are concerned about oversupply of cereals and low producer prices in many parts of the world, most African countries, except for South Africa, have grain deficits. Sub-Saharan Africa as a whole consumes (both directly and indirectly) 80 to 90 million tons of cereals per year, 20 percent of which are imported (Table 3.1). Grain net imports account for 15 and 25 percent of domestic consumption in East and Southern Africa, respectively, even though 40–50 percent of the regions' cultivated land is allocated to these crops. This relatively large share of food imports is not primarily a result of food aid, as that has constituted

[1]In this section East Africa includes Burundi, Comoros, Democratic Republic of Congo, Djibouti, Eritrea, Ethiopia, Kenya, Madagascar, Mauritius, Rwanda, Somalia, Tanzania, and Uganda. Southern Africa includes Angola, Botswana, Lesotho, Malawi, Mozambique, Namibia, Swaziland, Zambia, and Zimbabwe. South Africa is considered separately from other Southern Africa countries because of its economic size and development level. Western Africa includes Benin, Burkina Faso, Cameroon, Cape Verde, Central African Republic, Republic of Congo, Côte d'Ivoire, Ghana, Guinea, Guinea-Bissau, Liberia, Mali, Mauritania, Niger, Nigeria, São Tomé and Principe, Senegal, Sierra Leone, and Togo.

Table 3.1 Demand and supply of major cereal crops in Sub-Saharan Africa, 1998–2003 annual averages

	Production (million tons)	Imports (million tons)	Exports (million tons)	Consumption (million tons)	Per capita consumption (kilograms)	Share of imports in consumption (%)
East Africa						
Maize	10.4	0.7	0.1	11.0	43.0	5.5
Rice (milled equivalent)	2.6	0.6	0.0	3.2	12.1	18.8
Wheat	2.0	3.2	0.1	5.1	19.5	60.8
Other coarse grains	11.1	0.2	0.1	11.2	43.0	0.9
Total cereals	26.1	4.7	0.3	30.5	117.6	14.4
Southern Africa, excluding South Africa						
Maize	5.9	1.2	0.2	6.9	111.1	14.5
Rice (milled equivalent)	0.3	0.2	0.0	0.5	5.0	40.0
Wheat	0.4	1.1	0.1	1.4	19.6	71.4
Other coarse grains	0.9	0.2	0.0	1.1	16.0	18.2
Total cereals	7.4	2.6	0.3	9.7	151.7	23.7
South Africa						
Maize	9.0	0.4	0.9	8.5	188.9	–5.9
Rice (milled equivalent)	0.0	0.9	0.0	0.9	20.0	100.0
Wheat	2.2	0.6	0.2	2.6	57.8	15.4
Other coarse grains	0.5	0.0	0.1	0.4	8.9	–25.0
Total cereals	11.7	1.9	1.2	12.4	275.6	5.6
West Africa						
Maize	9.9	0.3	0.1	10.1	40.1	2.0
Rice (milled equivalent)	4.9	3.6	0.0	8.5	33.6	42.4
Wheat	0.1	4.0	0.1	4.0	15.9	97.5
Other coarse grains	22.8	0.3	0.0	23.1	93.3	1.3
Total cereals	37.7	8.2	0.2	45.7	182.9	17.5
Sub-Saharan Africa, excluding South Africa						
Maize	26.2	2.2	0.4	28.0	49.4	6.4
Rice (milled equivalent)	7.8	4.4	0.0	12.2	20.3	36.1
Wheat	2.5	8.3	0.3	10.5	17.8	76.2
Other coarse grains	34.8	0.7	0.1	35.4	60.6	1.7
Total cereals	71.2	15.5	0.8	85.9	148.1	17.1

Source: FAOSTAT (2006), last accessed in March 2006.

a relatively small and now declining share of total consumption (Stevens 2003). Rather, commercial imports of food are now a major drain on foreign exchange resources in many of these countries. In Sub-Saharan Africa, the cost of grain imports is equivalent to two-thirds of the revenues earned from the five leading traditional exports (cocoa, coffee, tea, tobacco, and cotton). Household level data show similar patterns. According to SIDA (2006), only about 10 percent of the households surveyed produced a marketable surplus of food, whereas more than half the households interviewed failed to produce enough food to cover their own needs and are therefore net buyers of basic food items.[2]

Consumption of livestock products accounts for 10–15 percent of African households' spending, but growth here is expected to be rapid, especially in poultry because of increased urbanization and per capita income. Currently, although Sub-Saharan Africa produces more than 6 million tons of meat and 16 million tons of milk, imports still account for 13 percent of total availability. For example, imports supply 17 percent of the poultry and 10 percent of the milk consumed in the region. The import share is particularly high in countries with rapid economic growth, such as Ghana and Uganda.

Growth in population and per capita food demand implies that food demand in the region is likely to increase considerably in the coming decade. According to a projection by FAO, Africa's population will increase from the 3-year average for 1997–1999 of 574 million to 883 million in 2015, with a growth rate of 2.6 percent per year (FAO 2002). During the same period, per capita GDP is expected to increase by 1.8 percent per year.[3] As a result, aggregate African demand for agricultural products will increase by 2.9 percent per year. In particular, according to the FAO projection, Africa's total demand for cereals will increase to 139 million metric tons by 2015, which is 62 percent higher than the average demand level between 1997 and 1999. Among the cereals, demand for maize alone will increase to 46 million tons by 2015, some 70 percent more than the current level. According to the same projection, demand for oilseeds and vegetable oils, meat, and dairy products will increase by 3.3, 3.4, and 2.9 percent per year, respectively.

IFPRI's IMPACT model projection suggests similar trends. Driven by population growth and urbanization, with a per capita income growth rate of 1 percent, a "business-as-usual" scenario in the model predicts that by 2020 Africa's total cereal demand is expected to be more than 50 percent higher and meat demand to be doubled from their current levels (Rosegrant et al. 2001). If the per capita income growth rate could be doubled—the rate required for achieving millennium development goals of halving the poverty and reducing hunger in Africa—the region's demand for livestock products would triple by 2020, given that the income elasticity of demand for livestock is high.

Although demand for cereals may not grow as fast as demand for livestock (since it is typically more income inelastic than livestock demand), increased feed demand will create market opportunities for maize and other coarse grains. According to the Chinese experience, a 2.5–3.0 of feed-to-meat ratio is needed for rapid development in the poultry industry. If Africa were to double its poultry production in the next 10 years (a 6 percent annual growth rate), feed grain demand would likewise rapidly increase. For example, if 30 percent of poultry products are produced using feed-grain-intensive production systems, feed demand by the poultry industry alone would be 1.5 to 2 million tons annually. There are also huge market opportunities for many staple crops and livestock products within Africa. Thus, growth in African agriculture and farmers' incomes could accelerate sharply if African agriculture becomes more competitive in its own regional and domestic markets (even if it is unable to compete in international markets for food products).

---

[2]The case studies of SIDA (2006) have been undertaken in eight African countries: Ethiopia, Ghana, Kenya, Malawi, Nigeria, Tanzania, Uganda, and Zambia. The village survey covered 3,000 smallholders in 103 villages.

[3]According to a World Bank study, per capita income in Africa as a whole will grow at 1.3 percent per year in the period to 2015 (World Bank 2004).

Exploring Regional Market Opportunities

Although national food demand is likely to grow rapidly in many countries and thus provide the most important marketing opportunities, a purely national perspective may not be the best one for agricultural growth strategies. Greater economic gains can be realized for groups of countries in various regions in Africa by improving marketing channels and coordinating them across borders. Through more integrated and competitive markets, individual countries can also specialize in the products in which they have a comparative advantage, thereby improving economic efficiency and unleashing regional growth dynamics.

Official statistics indicate that intraregional trade in Sub-Saharan Africa is minimal—only $2.0 billion per year (for details see Tables A2 and A5 for Sub-Saharan African exports and imports, and Tables A3, A4, A6, and A7, respectively, for East and Southern African exports and imports). Yet, there are striking differences in available resources among countries within Africa—more so than in most other areas—suggesting a potential for sizable regional trade. For example, countries with abundant land and scarce labor neighbor with others with abundant labor and scarce land. The presence of South Africa and other middle-income countries in Southern Africa provides the region with a unique opportunity to further exploit agricultural potential and regional trade through regional dynamics and integration (Nin Pratt and Diao 2006).

To help understand the potential intraregional demand among African countries, we examine the differences in comparative advantage across countries. As data on intraregional agricultural trade flows for food staples often significantly underreport actual trade, we compare instead indicators of "revealed" comparative advantage across countries. By matching commodities in the region for which some countries have a comparative advantage to export and others a comparative disadvantage (and have to import), we can get an indication of the potential for increases in regional trade where little is currently taking place (at least as indicated in the official statistics).

To begin with, we calculate a statistical indicator called the *"revealed" comparative advantage* (RCA), defined as the difference between a country's share of commodity *i*'s exports and imports relative to the share of this commodity in the world total trade, that is,

$$\mathrm{RCA}_i = \frac{\frac{x_i}{X} - \frac{m_i}{M}}{\frac{x_i^w}{X^w}},$$

where $\frac{x_i}{X}$ and $\frac{m_i}{M}$ are the shares of commodity *i*'s exports and imports in a country's total exports and imports, respectively, and $\frac{x_i^w}{X^w}$ is the share of commodity *i*'s trade in world total trade (since world exports equal imports for each commodity). A country is said to have a comparative advantage (disadvantage) in commodity *i* if the value of RCA_i is positive (negative; Balassa 1965).[4] Note, though, that this measure ignores any trade policy distortions that affect trade flows. The calculated RCA indices show that African countries have different comparative advantages in a wide range of diversified agricultural commodities. At five-digit SITC classification, there are more than 200 agricultural commodities in which there are only one to three relatively large exporting countries that have a strong com-

[4]We use five-digit SITC trade data drawn from the United Nations COMTRADE database for the calculation of RCA. The database includes 33 African countries: Angola, Benin, Botswana, Burkina Faso, Cameroon, Chad, Democratic Republic of Congo, Côte d'Ivoire, Ethiopia, Gambia, Ghana, Kenya, Lesotho, Madagascar, Malawi, Mali, Mauritania, Mozambique, Namibia, Niger, Nigeria, Rwanda, Senegal, Sierra Leone, Somalia, South Africa, Sudan, Swaziland, Tanzania, Togo, Uganda, Zambia, and Zimbabwe. See Diao and Yanoma (2003) for the calculation of RCA in detail.

Table 3.2 Correlation between exportables and importables for the major trading countries in Sub-Saharan Africa

	Total agriculture	Food staples	Fruits and vegetables
1990–95			
Number of commodities			
RCA index > 4.0	260	58	38
RCA index < –4.0	198	63	17
Number of commodities appearing in both	83	28	7
Correlation coefficient	0.595[a]	0.341[a]	–0.135[a]
1996–2000			
Number of commodities	299	61	43
RCA index > 4.0	212	69	20
RCA index < –4.0			
Number of commodities appearing in both	97	29	9
Correlation coefficient	0.624[a]	0.629[a]	0.078

Source: Authors' calculation using data from UNCOMTRADE 2005, last accessed in April 2005.
[a]Correlation is significant at 99% (two-tailed) run by SPSS.

parative advantage (in the relevant time period 1990–2000).[5]

We then look at the match between the commodities in which exporting countries have a strong comparative advantage and those imported by other countries in the region. We utilize two commodity lists—one in which the RCA values are high and positive (greater than 4.0) for the large exporters, and the other in which the RCA values are low and negative (less than –4.0) for the large importers—to estimate the correlation between the two groups of products. As in any other correlation analysis, the coefficient values lie between –1.0 and +1.0, and a value close to unity implies that the two commodity groups, exportables and importables, are highly correlated. If exportable commodities are quite different from importable ones, the correlation coefficient would be significantly less than unity or perhaps even negative.

Table 3.2 presents the results of this analysis for 1990–95 and 1996–2000. In the period from 1990 to 1995 some 83 commodities appear in both the export and import lists; this number accounts for slightly less than one-third of the commodities (260) in which at least one large exporting country has strong comparative advantage (with a high and positive RCA value) and less than half of the commodities (198) in which at least one large importing country has strong comparative disadvantage (with a low and negative RCA value). The estimated correlation coefficient between the lists for exports (260) and imports (198) is 0.595. From 1996 to 2000, the number of commodities that appear in both lists and the value of correlation coefficient both increase (Table 3.2, first column).

These results indicate a potential for strengthening mutually beneficial agricultural trade in Africa. Moreover, the appar-

[5]We define those countries, including South Africa, with more than 2 percent of total African agricultural exports (imports) or more than 10 percent of intraregional exports (imports) in 1996–2000 as large exporting (importing) countries.

ently slightly increased value of the coefficient and increased number of commodities appearing on both lists in the second period seem to show an increased trend in such potential.

Results regarding the potential in intraregional trade of staple foods are likewise encouraging. As discussed earlier, official data indicate that intraregional trade in these commodities is limited at present, with most imports of food staples coming from outside the region. In the first period, there are 58 staple commodities for which the RCA value is high and 63 staple commodities for which it is low. These commodities include livestock and livestock products, fishery, cereals, roots and tubers, pulses, and oilseeds. Twenty-eight staple commodities appear in both lists. The correlation coefficient of the exportable and importable staple commodities in the first period is 0.341. In the second period, the number of commodities that appear in the exportable (importable) lists increases to 61 (69), with 29 appearing in both lists. The value of the estimated correlation coefficient rises to 0.629, almost double its value in the first period (Table 3.2, second column). This result indicates that foodstuffs are among the most dynamic products in potential intra-African trade. Our findings are supported by Yeats (1998), who reports that feedstuff accounts for the fastest growing products in intraregional trade.

Our findings are also supported by the recent evolution of regional trade, especially in Southern Africa. According to Nin Pratt and Diao (2006), regional trade in Southern Africa has expanded significantly in the last 10 years, and the growth rate of intraregional agricultural trade was almost twice that of the total of agricultural exports of the region, though starting from a smaller base. Such expansion of regional trade is associated with South Africa's increasing involvement since the country was readmitted to the global economy community in 1994. Since then, South Africa has been an active player in Southern Africa's commodity and capital markets, as 25 percent of the region's total foreign direct investments are from South Africa and 75 percent of its export expansion can be accounted for by increased exports from South Africa.

We further checked whether there is a correlation of revealed comparative advantage for trade in fruits and vegetables. We find that the estimated correlation coefficient is negative (–0.14) in the first period and becomes positive but small (0.08) in the second period (Table 3.2, third column), but the estimates are not statistically significant in either period. This outcome seems to show that there is a poor match in fruit and vegetable trade between intra-African exports and imports and that the most promising opportunities for rapid growth of nontraditional exports is outside the region. Nonetheless, even though countries within Africa are mainly competitors in this trade, regional linkages may come through technological spillover and learning from each other.

Constraints to Exploring Market Opportunities

Improving competitiveness of agricultural products in international, regional, and domestic markets is the key to expanding market opportunities. As outlined earlier, there is great potential for African agricultural products, especially in the regional and domestic markets, but Africa's agricultural products are often not competitive. Low productivity, high transportation and marketing costs, various formal and informal trade barriers (both physical and institutional), and inconsistencies in trade and agricultural policies among African countries all contribute to the lack of market competitiveness. In order to be more competitive, the region must increase overall agricultural productivity, reform infrastructure and institutions to reduce transportation and transaction costs, reform policies to encourage agricultural trade among the countries in the region, and enhance re-

Table 3.3 Ratio of wholesale maize prices to import parity price, 1994–2002

	Uganda	Kenya	Zambia	Mozambique	Tanzania
1994	1.23	1.5	0.91		
1995	0.88	1.03	1.01		1.07
1996	1.04	0.88	0.75		1.04
1997	1.96	1.64	0.93		1.44
1998	1.54	1.37	1.03	1.9	1.5
1999	1.36	1.58	0.89	1.18	2.02
2000	1.08	1.68		1.35	
2001			1.52	1.67	
2002			2.82	2.02	

Source: Authors' calculations, using data from various sources.

gional cooperation to open European and American markets by removing asymmetric agricultural and trade policies.

The case of maize provides one example of the extent of relative competitiveness of African agricultural products in the world and regional markets. As shown in Table 3.3, the ratio of annual average wholesale to import parity prices for maize in the international market in this group of countries was greater than one, indicating that they are not competitive in the international maize market.[6] However, some African countries may have relative competitiveness in regional markets. Significant price differentials among African countries seem to suggest potential for intraregional maize trade, particularly in years of short supply (Dréze and Sen 1989; Dorosh 2001). Nonetheless, in some years, grain from outside the region may be less costly than imports from neighbors, as in Kenya and Uganda from 1997 to 2000 (Figure 3.1).

Improvements in agricultural productivity, though important, are not likely to make agricultural products competitive in international markets without substantial reductions in transport and marketing costs. Transport, communication, and marketing infrastructure in Africa are rudimentary, even by historical standards compared with other developing countries. For example, in the early 1990s Sub-Saharan Africa had about one-sixth of the rural road density of India in 1950. Poor infrastructure and poorly functioning domestic and regional markets add enormous transportation and marketing costs to production costs and squeeze African traders and farmers out of their domestic and regional markets. It is not uncommon for African farmers to receive only 10 to 20 percent of the price share from the products they sell with the rest—80 to 90 percent—being the costs of transportation and marketing.

Many African countries have undertaken market reforms in recent years.

[6]The import price is the U.S. FOB yellow maize price plus $45.4 per ton in shipping and other marketing costs to South Africa's ports. This price also gives an approximation of import parity in other ports in Sub-Saharan Africa. Given additional transport and marketing costs required to bring maize from wholesale markets in the sample countries to ports, the figures show that domestically produced maize is generally not competitive with imported maize.

Figure 3.1 Maize wholesale prices in Uganda and Kenya and import parity price

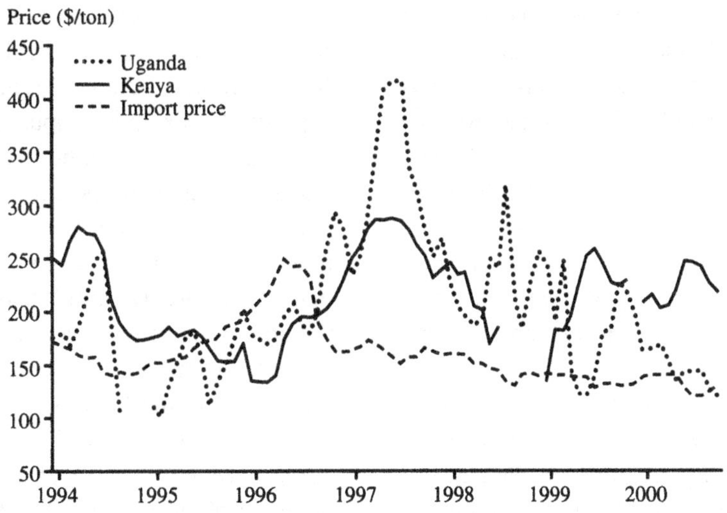

Source: Authors' calculation using data from various sources.

These reforms are a solid beginning but are not yet sufficient to generate greater supply response and competitiveness in the export, domestic, and regional markets (Kherallah et al. 2002). Although market liberalization may have removed price distortions, reduced marketing margins, and improved market integration, a recent study by SIDA (2006) shows that only large commercial farmers in Africa have found opportunities to diversify, whereas smallholders have been progressively marginalized from, rather than integrated into, the liberalized market. Agricultural markets, both for inputs and outputs, remain underdeveloped. Small-scale farmers in food crop production, especially those living far from roads and markets, are not only unable to benefit from the liberalization, but may no longer be able to afford chemical fertilizer owing to increased input prices. In the domestic and regional food product markets, such as those for cereal and livestock products, the role of market institutions in supporting exchange is especially weak. Public policies often hinder the informal trade and small-scale traders who are playing critical roles in the regional and domestic markets. Marketing margins for staple food trade are high and farmers' share in the final prices is low, but value added to traders is often barely beyond transport costs.

It is urgent for African countries to design appropriate market development policies to address these issues. Gabre-Madhin (2001), for example, studied market institutions, transaction costs, and social capital in the Ethiopian grain market and suggested three areas of policy intervention that are relevant to other African countries' grain and other staple food markets in which small traders play a dominant role. They include strengthening institutions responsible for standards, quality control, credit and other financial supports; contract enforcement; and market information dissemination. The case studies by Minot and Ngigi (2003) on the Kenyan horticultural sector show that through various self-help growers' groups, collective marketing helps small farmers make contractual market arrangements that offer a more reliable market outlet, allows them to obtain the seed on credit, and provides them with other production advice.

In the growing competitive export and domestic markets, African farmers also need instruments to help them manage price and market risks, such as efficient, targeted safety-net programs and market-based risk-management vehicles, for example, weather insurance and futures price contracts (Skees, Hazell, and Miranda 1999). At the same time, for farmers to become better integrated into today's increasingly competitive markets, they must acquire the technologies they need to improve quality and postharvest storage and processing.

Agricultural commodity flows in Africa are also severely hampered by both formal and informal intraregional trade barriers. Without better regional integration through lowering and removal of these barriers between one African country and another, harmonization of regulations to promote cross-border investment, and creation of institutional frameworks such as the Common Market for Eastern and Southern Africa (COMESA), the Southern Africa Development Community (SADC), and the Southern Africa Custom Union (SACU) to encourage free movement of people, commodities, and capital, improvement in transportation conditions within a country may not necessarily increase intraregional trade and may actually lead to an increase in extraregional imports.

South Africa is a potential market for its neighbors' agricultural exports and they, in turn, represent a potentially large market for South Africa's nonagricultural and high-value agricultural exports. However, the high tariffs imposed by South Africa on imported dairy products and cereals and those imposed by the low-income Southern African countries on imports of textiles, fruits, vegetables, and processed food precuts from South Africa hinder the development of an integrated regional market. Increasing agricultural trade within Africa is also constrained by the trade barriers and agricultural policies in developed countries, as many of them, such as countries in the Organization for Economic Cooperation and Development (OECD), are the main players in the world market. Most OECD countries have traditionally afforded heavy protection for their agricultural sectors, partly through policies granting domestic support (for example, input subsidies) and partly through trade policies such as tariffs, quotas, export subsidies, and seasonal import restrictions. These policies not only make African agricultural products uncompetitive in OECD markets, but impact on the domestic and regional markets in Africa as well. Many studies show that policies pursued by a small number of OECD countries cause most of the distortions in world agricultural markets (see, for example, Burfisher, Robinson, and Thierfelder 2002; Diao, Roe, and Somwaru 2002). By decomposing the possible increase in world prices owing to worldwide agricultural trade liberalization, Diao, Roe, and Somwaru (2002) find that removing domestic and export subsidies for agriculture in more developed countries, especially in the European Union and the United States, would account for about 80 percent of the possible increase in world agricultural prices after worldwide agricultural liberalization. Research at IFPRI suggests that global agricultural trade liberalization would benefit the whole of Africa by at least $5 billion per year (Diao and Yanoma 2003), and this figure does not take into account the potential gains from future agricultural investment stimulated by better prices for farmers.

Summary

In contrast to most traditional and nontraditional agricultural exports, domestic demand and production of food crops and livestock products far exceed internationally traded quantities in Sub-Saharan Africa. Supply of many staple commodities cannot meet the current demand, and 20 percent of the region's demand for food grains such as maize, rice, and wheat is met by imports from abroad. Displacing some of these food imports with domestic supply could have an enormous impact on overall African farm income and agricultural growth in the region. Sub-Saharan Africa's total food demand is projected to further increase considerably in the next 10 years (for example, cereal demand will be 60 percent higher than current levels). Existing and growing food demand in the region offers great potential for significant expansion of markets for many commonly grown cereal, root, tuber, pulse, oilseed, and livestock products. Nevertheless, the challenge is to satisfy current and future food demand in Africa with domestic and regional production rather than by imports from abroad.

The regional comparative advantage analysis shows that many African countries import a long list of agricultural commodities from outside the region, even though their regional neighbors have comparative advantages in producing them. In particular, the analysis shows that the greatest growth potential in intraregional trade lies with food staples. However, actual intraregional trade of food staples is at only a small fraction of its potential as African products fail to compete with subsidized low-cost food imports from wealthy exporting countries. Apart from the asymmetric agricultural and trade policies in developed countries, low productivity, high transportation and marketing costs, trade restrictions, and domestic pricing policies in African countries are major constraints for intraregional trade.

There is scope for integrating and expanding domestic and regional markets for a range of key commodities, but to explore the opportunities, African countries have to improve their competitiveness in agricultural production and marketing. The region as a whole must increase overall agricultural productivity, reform infrastructure and institutions to reduce transportation and transaction costs, reform policies to encourage intraregional agricultural trade, and enhance regional cooperation to put political pressure on developed countries (primarily in Europe and North America) to remove asymmetric agricultural and trade policies (in particular, domestic and export subsidies).

Given prevailing world agricultural trade policies and poorly functioning markets within Africa, rapid increases in agricultural production might lead to price declines in the short run. However, in the long run, efforts to reduce marketing (transportation and transaction) costs can go a long way toward reducing consumer food prices while raising producer incomes. By investing in infrastructure and institutions, African countries can maximize the positive linkage effects of growth in farm and nonfarm products. Over time, these linkages will significantly increase gains in overall incomes and calorie consumption, loosening the grip of hunger on the African continent.

CHAPTER 4

Economywide Analysis of Agricultural Growth Opportunities

In this chapter we employ a world general equilibrium model to assess three broad alternative agricultural development strategies (promoting traditional export, nontraditional export, and food staples) in East and Southern Africa for overall economic growth. The potential for investment and higher productivity in agriculture to increase income and food consumption depends on the diverse economic structures across countries, the initial size of the sector, the magnitude of the change in the sector's productivity, and the structure of production and demand within each country. There are also potentially important linkages across sectors and countries through changes in relative prices, including changes in the real exchange rate, and adjustment in resource allocation at sector and country levels. Our analysis focuses mainly on demand-side constraints. The simulations assume exogenous technological change in the production of various crops which results in increases in exports of traditional and nontraditional agricultural products, changing real market prices, total farm income, and national income.

Methodology

Our analysis of alternative growth scenarios for African agriculture focuses on East and Southern Africa. The world general equilibrium model developed for the analysis includes nine countries in East and Southern Africa: Botswana, Madagascar, Malawi, Mozambique, Republic of South Africa, Tanzania, Uganda, Zambia, and Zimbabwe, as well as two aggregate subregions, the Southern African Custom Union (SACU)[1] and the rest of Southern Africa. West Africa is aggregated into a region called "the rest of Sub-Saharan Africa region."[2] The rest of the world is modeled as ten other countries/regions.[3] Our study focuses on the seven low-income countries in East and Southern Africa. Table A8 highlights the economic structure of these seven countries, which provides the initial conditions for the modeling analysis.

We model 34 production sectors, 23 of which are agricultural or agriculture-related (see Table B1 for commodity/sector details). Commodities produced by production sectors are ei-

[1]SACU member countries are Botswana, Lesotho, Namibia, South Africa, and Swaziland.

[2]Data are from the Global Trade Analysis Project (GTAP) database version 6 (2005) for 2001. Individual West African countries are not listed in the database, which precludes a disaggregated analysis. Similarly, because Kenya is not shown as a separate country, but is included in the "rest of the Sub-Saharan Africa" region, we are unable to present country-specific results for Kenya.

[3]The other countries/regions specified can be found in Table B1.

ther sold to the domestic markets or exported to other countries/regions. World prices adjust to clear world export/import markets for each commodity. International trade is traced by import destination and export sources; that is, the bilateral and multilateral trade flows among countries/regions, rather than net trade or aggregate exports and imports, are included in the model. Thus, both intra-African regional and interregional trade (for the nine countries included in the model) is taken into account.

Production technology is represented by fixed input-output coefficients for intermediate goods (such as fertilizers) and a constant elasticity substitution (CES) function for primary inputs: two types of labor (skilled and unskilled), land, other natural resources, and capital. These production technologies vary across sectors and countries, to capture the differential initial economic structure specified in the base social accounting matrices for the countries/regions in the model.

Consumer income levels are determined as the sum of factor incomes less direct taxes. Consumption demand is modeled as a linear expenditure system (LES) demand function for each commodity. The income elasticities used to derive the marginal budget shares for consumption are from Reimer and Hertel (2004), in which, for example, income elasticity of demand for grain is 0.4–0.5 for the low-income African countries. We assume a single Frisch parameter for each individual country/region.[4] The price elasticities for grain demand, the key parameters in these simulations, range from −0.15 to −0.34.

The model assumes imperfect substitution between domestic and foreign goods, as well as imperfect substitution across goods produced in each country. The degree of price transmission between domestic and international markets for each good or service is thus determined by the elasticities of substitution between local and internationally traded goods and services, and the share of traded goods or services in total supply or demand. Domestic prices of goods and services do not track world prices closely unless the share of exports (imports) in domestic production (consumption) is large. The mathematical description of the model is detailed in Appendix B.

Simulation Results

In order to evaluate possible market opportunities or constraints related to agricultural growth in the sample countries, we examine six main growth scenarios. The first four model the effects of an exogenous increase in total factor productivity (TFP) in the traditional agricultural export sector (Scenario 1), the nontraditional export sector (Scenario 2), the grain sector (Scenario 3),[5] and the grain and livestock sectors combined (Scenario 4). To further explore the effects of removing market constraints on the growth of the grain sector, we also simulate a grain trading facility in Scenario 3a. The last two scenarios combine the effects of growth in agriculture with a reduction in marketing costs through an increase in the transport sector's productivity (Scenario 5) and increases in nonagricultural growth in selected sectors (Scenario 6) to further capture the agricultural and nonagricultural linkages in growth. For each scenario, we use two different assumptions about labor supply, full employment and unemployment in unskilled labor, which implies that there are two different sets of general equilibrium solutions for each of the six scenarios. Most of the discussions in this section are

[4]See Dervis, de Melo, and Robinson (1982), pp. 482–483, for a discussion of the Frisch parameter and formulas to calculate price elasticities.

[5]We focus the analysis on grains, since the GTAP database used does not separate out roots and tubers (cassava is an important staple in parts of Mozambique, Zambia, and other countries in East and Southern Africa).

Table 4.1 Description of the model scenarios

Scenario	Sector TFP annual growth rate (%)
Scenario 1: High growth of traditional exports	
Cotton	4.5
Other exportable crops, including tree crops	3.2
Scenario 2: High growth of nontraditional exports	
Exportable vegetables and fruits	6.0
Scenario 3: High growth of grain sector	
Rice, wheat, maize, and other coarse grains	2.5
Scenario 4: High growth of grains and livestock	
Grains	2.5
Livestock[a]	3.0
Scenario 5: High growth in agriculture (Scenarios 1, 2, and 4) plus increased productivity of marketing	
Transport sector	4.5
Scenario 6: With high growth in selected nonagricultural sectors (Scenario 5 plus increased productivity of five nonagricultural sectors)	
Selected manufacturing and private services	3.0

[a]Capital stock increased by the same percentage as total factor productivity (TFP); we also model an exogenous increase in consumer preference for livestock products.

based on the full-employment assumption, although, if necessary, the equilibrium results with the assumption of unemployed unskilled labor are discussed in comparison with the results with the full-employment assumption. Table 4.1 summarizes other principal assumptions of each scenario.

Scenario 1: Recovery of the Market Share of Traditional Export Crops

In Scenario 1, we simulate a recovery of lost markets for Africa's traditional agricultural exports, including coffee, cocoa, cotton, tea, and tobacco.[6] In this simulation, we increase the TFP of traditional agricultural export sectors by 4.5 percent for cotton and 3.2 percent for other traditional exports for 12 years for the seven low-income countries in East and Southern Africa.[7]

This optimistic scenario for traditional export crops produces somewhat disappointing results in terms of the growth of real GDP, however. With a fixed-labor-supply assumption, the 3–5 percent TFP annual growth rate in the traditional export sectors modeled here increases real GDP per capita in the countries under study by 0.13 percent on average annually and per capita total agricultural real income by 0.48 percent (Table A9, part A, row 1). In alternative simulations that assume the existence of unemployed unskilled labor, additional per capita GDP growth rises to 0.28 percent

[6]Tobacco, coffee, cocoa, and tea are aggregated into the "exportable other crops" sector in the GTAP database.

[7]A TFP growth rate of 3.2 percent is chosen such that the region returns to its historically highest level of such exports.

and agricultural GDP to 0.67 percent per year (Table A9, part B, row 1). At the individual country level, additional annual growth in a country's per capita GDP ranges from as low as 0.08–0.11 percent in Mozambique, Malawi, Zambia, and Zimbabwe to 0.14–0.15 percent in Madagascar, Uganda, and Tanzania assuming fixed labor supply, and 0.16–0.47 percent in the seven countries assuming underemployed unskilled labor (Table A9, column 4).

The 3 to 5 percent TFP annual growth rate in the traditional export sectors modeled here results in different sectoral production and export growth owing to the different sizes of each sector in the sample countries. For example, tobacco is a dominant traditional export crop in Malawi and Zimbabwe, but in the simulation annual growth is only 1.9 percent and 3.2 percent, respectively, in traditional exportable production. On the other hand, the sector's production grows at 5–9 percent in the other five low-income countries. Increases in production result in rapid growth in traditional exports, a growth rate higher than 10 percent for five countries, all except Malawi and Zimbabwe, in which a much more modest growth rate (1.9 and 3.6 percent, respectively) is observed. However, given that the size of the traditional agricultural export sector is relatively large in Malawi and Zimbabwe (13 and 29 percent, respectively, in the data), growth in total agricultural exports is comparable across the seven countries, ranging from 1.37 percent in Malawi to 4.71 percent in Uganda (Table A9, part A, column 9).[8]

There are two major reasons why rapid growth in traditional exports fails to increase real GDP substantially in this scenario. First, for most of the sample countries, the traditional export sector accounts for a small share of the total agricultural GDP, even though it accounts for a large share in its total agricultural exports. Moreover, this simulation does not assume a return of world prices to historical high levels. In fact, in the simulation, increased production and exports cause the world prices for the traditional commodities to fall slightly. An increase in world prices to the average levels of the mid-1970s (historical peak period for many crops) would increase the revenues generated from exports, but such a recovery in medium-term average prices seems unlikely.

The second reason is that markets for the traditional export commodities involve relatively little domestic processing and therefore generate only weak linkage effects through the domestic market economy. Only the farmers cultivating these crops reap major direct benefits from increased exports, and their increased income and that from the directly related nonagricultural sectors cannot provide enough demand to stimulate further growth in broad agricultural and nonagricultural sectors. Calculated growth multipliers (real GDP growth induced by a one unit increase in the traditional export sector's value added) are quite small for most countries (see Table 4.3, column 1). Note that market opportunities and hence growth contributions of traditional agricultural products can significantly increase if the quality of the products can be improved or if the products can be processed (even just go through primary processing) within the region. There is a substantial difference between a brand product, for example, a gourmet coffee, and a nonbrand product. Improving product quality will generate more farm income from the same volume of production, and processing the products in the region will provide more employment and improve the linkage effects both within the producing country and to some extent to neighboring African trading partners.

[8]Cotton is actually a nontraditional export crop in Zambia. We have to treat it as a traditional sector in all countries in order to compare its economywide effect across countries.

Table 4.2 Increased per capita GDP growth rate in model simulations (additional annual %)

	Sim 1: Growth in traditional exports	Sim 2: Growth in nontraditional exports	Sim 3: Growth in grains	Sim 3a: Growth in grains with trade facility	Sim 4: Growth in grains and livestock	Combining simulations 1, 2, and 4	Sim 5: Combining simulations 1, 2, and 4 with growth in transport sector	Sim 6: Simulation 5 with growth in selected nonagricultural sectors
A. With fixed labor supply								
Malawi	0.09	0.03	0.19	0.23	0.35	0.47	1.13	2.40
Madagascar	0.14	0.03	0.18	0.18	0.64	0.82	1.24	1.92
Mozambique	0.08	0.02	0.11	0.20	0.22	0.33	0.62	1.77
Tanzania	0.15	0.05	0.27	0.26	0.54	0.76	1.13	1.98
Uganda	0.14	0.02	0.15	0.18	0.49	0.66	0.95	1.44
Zambia	0.10	0.01	0.09	0.11	0.19	0.32	1.06	2.12
Zimbabwe	0.11	0.01	0.03	0.04	0.16	0.27	0.73	1.67
B. With unemployed unskilled labor								
Malawi	0.16	0.06	0.22	0.39	0.49	0.70	1.59	3.47
Madagascar	0.34	0.08	0.21	0.24	1.05	1.51	2.23	3.32
Mozambique	0.17	0.06	0.16	0.71	0.34	0.60	1.06	2.91
Tanzania	0.33	0.11	0.32	0.25	0.72	1.22	1.89	3.44
Uganda	0.47	0.11	0.23	0.33	0.81	1.44	2.11	3.11
Zambia	0.21	0.03	0.10	0.20	0.27	0.54	1.48	3.06
Zimbabwe	0.16	0.02	0.03	0.03	0.23	0.39	0.99	2.31

Sources: CGE model results.

Scenario 2: Rapid Growth of Nontraditional Agricultural Exports

In Scenario 2, we model an increase in nontraditional agricultural exports by increasing the TFP of the nontraditional agricultural export sector by 4 percent annually for 12 years and endogenously increasing capital stock in the sector by 3.2–7.9 percent, except for Zimbabwe, such that the growth rate in the sample countries' exportable fruit and vegetable sector production approximates the growth rate that Kenya has experienced in its horticultural sector in the last 20 years.

In this scenario, in which they face no significant market constraints,[9] nontraditional exports increase annually by 16 percent in Madagascar to 24 percent in Uganda. In Zimbabwe, the increase is only 10 percent per year because of a much smaller increase in capital accumulation (0.9 percent per year). These growth rates almost double the average annual growth rate of Kenya's fruit and vegetable exports over the past 20 years.[10]

However, assuming a fixed labor supply, accelerated growth in nontraditional exports raises these countries' per capita real GDP by only 0.01–0.05 percent per year (Table 4.2, part A, column 2). The growth rate of agricultural income is also modest: 0.21 percent per capita in Mozambique, 0.10–0.16 percent in Madagascar, Malawi, Tanzania, and Uganda, with slight declines in Zambia and Zimbabwe. If unemployment in unskilled labor is assumed,[11] high growth in nontraditional exports would only create additional annual employment opportunities equivalent to 0.09–0.29 percent of the labor forces that are currently employed in the agricultural sector in the six countries (and 0.06 percent in Zimbabwe). More employment opportunities enhance the effect of economywide linkages of nontraditional exports, resulting in an annual growth rate of real GDP that is more than double in each of the seven countries under the unemployed-unskilled-labor assumption. In Malawi, Madagascar, Mozambique, Tanzania, and Uganda, for example, the per capita annual growth rate of real GDP rises to 0.06–0.11 percent per year with unemployed unskilled labor, from 0.02–0.05 percent with a fixed labor supply (Table 4.2, part B, column 2).

Thus, these simulations show that given the small initial size of the sector, even a very rapid rate of increase in exports does not contribute much to growth in total agricultural income or total GDP for the countries included in the sample. Note, however, that this simulation may underestimate the dynamics of the sector and its growth linkage effect through induced technological change and investment in nonagriculture, such as in processing, packaging, transporting, and marketing sectors, given that the model does not include any endogenous change in technology or investment induced by growth in nontraditional exports. Experiences with both nontraditional agricultural exports in Kenya and export-led growth in East Asia have shown much stronger dynamic linkages between exports and technological changes and investment (both in agriculture and nonagriculture). We try to partially address these issues in Scenarios 5 and 6.

Scenario 3: Productivity Growth in the Grain Sector

In Scenario 3, we increase the TFP in the cereal sector (rice, wheat, maize, and other

[9]In the model simulations, increased nontraditional exports by East and Southern African low-income countries do not lead to a decline in world prices because of the region's very small initial share in total world trade in these products.

[10]These growth rates in exports are derived assuming fixed labor supplies with no unemployment and market-clearing real wages in each country.

[11]To simulate potential growth linkages due to increased productive use of previously underemployed factors, these simulations assume a fixed real wage and endogenous labor supply.

Table 4.3 Growth multipliers in simulations (real GDP growth induced by one unit of increase in shocked sector's value added, both in base-year prices)

	Sim 1: Growth in traditional exports	Sim 2: Growth in nontraditional exports	Sim 3: Growth in grains	Sim 3a: Growth in grains with trade facility	Sim 4: Growth in grains and livestock	Combining simulations 1, 2, and 4	Sim 5: Combining simulations 1, 2, and 4 with growth in transport sector	Sim 6: Simulation 5 with growth in selected nonagricultural sectors
A. With fixed labor supply								
Malawi	1.06	0.59	0.85	0.75	0.81	0.83	1.00	1.10
Madagascar	0.68	0.63	2.92	1.68	0.88	0.89	1.07	1.19
Mozambique	0.48	0.69	1.02	0.41	0.72	0.69	1.06	1.22
Tanzania	0.65	0.61	1.52	0.93	0.90	0.91	1.10	1.20
Uganda	0.44	0.49	1.15	0.50	0.70	0.73	0.88	1.03
Zambia	0.76	0.51	1.14	0.82	0.74	0.74	1.27	1.19
Zimbabwe	1.12	0.82	1.63	2.07	0.71	0.91	1.24	1.25
B. With unemployed unskilled labor								
Malawi	1.79	1.16	1.80	1.20	1.08	1.15	1.23	1.18
Madagascar	1.49	1.83	3.47	2.19	1.29	1.32	1.41	1.34
Mozambique	0.99	2.50	1.88	1.39	1.08	1.17	1.55	1.44
Tanzania	1.36	1.18	2.38	0.89	1.13	1.24	1.43	1.39
Uganda	1.35	2.63	1.86	0.90	1.06	1.27	1.42	1.47
Zambia	1.48	1.04	2.97	1.49	1.00	1.17	1.52	1.30
Zimbabwe	1.59	1.15	2.89	1.73	1.02	1.26	1.51	1.35

cereals) by 2.5 percent per year, which results in additional growth in grain production by 0.55 percent in Zimbabwe to 2.13 percent in Mozambique annually (Table A10, part A, column 1). The real prices of grains fall, since the increase in supply exceeds the growth in demand. As a result, farm income falls in five of the seven countries and rises slightly in Mozambique and Uganda (Table A10, part A, column 6). Consumers as a whole benefit from increased overall growth and lower grain prices, and food consumption increases by 0.11–0.43 percent annually in the seven countries (Table A10, part A, column 7). Moreover, in addition to the 0.23–1.06 percent increase in consumption of grain products, demand for other food commodities also increases. Overall, growth in per capita real GDP increases by 0.09 to 0.27 percent per year in six countries, but only 0.03 percent in Zimbabwe, owing to adverse price impacts, as well as to the relatively small share of maize production in overall GDP (Table A10, part A, column 5).

Differences in the magnitude of the initial increases in sectoral production to some extent obscure comparisons of the efficiency of the alternative growth strategies presented earlier. To facilitate the comparison of alternative investments and strategies, we calculate growth multipliers defined as the total increase in real GDP divided by the increase in the shocked sector's total value added, both measured at the initial (base-year) level of prices.[12]

As shown in Table 4.3, the grain sector's growth multipliers are in general large. A one unit increase in the grain sector's output (in terms of value added measured by the base-year prices) results in 1.80–3.47 unit increase in total real GDP across the seven countries (assuming underemployed unskilled labor; Table 4.3, part B, column 3). These grain sector multipliers are higher than multipliers in the two exportable agricultural subsectors in most countries. The exceptions are Mozambique and Uganda, in which the nontraditional export sector has the highest growth multiplier. Nontraditional agricultural export growth multipliers are 2.50 and 2.63 in Mozambique and Uganda, respectively, as compared to 1.04–1.83 in the other five countries. The traditional export growth multiplier is close to or greater than 1.5 in four countries, but less than 1 in Mozambique (Table 4.3, part B, columns 1–3). Moreover, productivity growth in food crop sectors benefits consumers through reductions in real prices of food, in contrast to productivity growth in export sectors that generates positive impacts on nonproducing households only through income multiplier effects.[13]

Scenario 3a: Productivity Growth in the Grain Sector with a Grain Trading Facility

One possible mechanism to help maintain the domestic agricultural terms of trade is through government intervention in markets to increase import substitution and total demand. Following Adelman (1984), we model a grain-trading facility in Scenario 3a that buys up part of the domestic supply. The cost of these grain purchases is covered by a lump sum tax on the consumers as a group.[14]

Compared with Scenario 3 in which grain imports fall by only 4.24–8.78 percent in the six countries (and fall by 2.45 percent

[12]See Bell and Hazell (1980) for an early methodological discussion of alternative multiplier models used in growth linkage analysis, and the discussion of Haggblade, Hammer, and Hazell (1991) on the improvement in the multiplier models with limited price endogeneity.

[13]The resulting multipliers derived using CGE models are in general relatively smaller than the standard fixed-price multipliers.

[14]In the model simulation, purchase of the grain-trading facility is a new endogenous variable that is perfectly substituted with imported grains.

in Zambia), introducing the grain trading facility results in a decline of grain imports by 22.5–51.8 percent (Table A10, part B, column 3). As a result of a smaller (in absolute magnitude) negative terms-of-trade effect on grain producers, grain production further increases in all seven countries (compared to the simulations with the same level of TFP growth rate). In general, the production increases are larger in the countries with relatively large declines in imports. For example, the growth rate of grain production rises to 12.1 percent in Mozambique (from 2.1 percent), and imports of grain fall by 65.4 percent. Similarly, grain production grows at rates of 6.3 and 3.6 percent, respectively, in Uganda and Malawi (compared to 1.9 and 1.4 percent in Scenario 3), and imports fall by 41.6 and 51.8 percent. As a result of this import substitution in Scenario 3a, per capita real GDP annual growth rises by 0.04 to 0.26 percent in the seven countries (Table 4.2, part A, column 4).

Scenario 4: Productivity Growth in the Grain and Livestock Sectors

Another way to improve terms of trade in the grain sector is to have simultaneous growth in livestock production. Thus, in Scenario 4, in addition to increasing TFP growth in the grain sector by 2.5 percent (Scenario 3), we increase TFP and the capital stock level in the livestock sector by 3 percent per year.[15] We also adjust consumer preferences toward demand for more livestock products at the same price and income levels, which reflects long-run changes in consumer tastes and livestock product availability not captured with income and price elasticities.

In this simulation, grain production does not increase in most countries relative to Scenario 3, and declines in grain prices become relatively modest. Thus, per capita GDP growth rates relative to those in Scenario 3 (increased productivity in grain sectors alone) are close to double or even more than double in all seven countries. For example, the growth rate of per capita GDP increases to 0.64 percent (from 0.18 percent in Scenario 3) in Madagascar, and to 0.54 and 0.49 percent in Tanzania and Uganda, respectively (Table 4.2, part A, column 5). The more positive effect on GDP in this scenario is largely due to productivity gains in livestock, as well as greater demand for grain in livestock feed and as food, which results in grain prices falling less than in Scenario 3. Consumers also have much greater benefit than in the grain-growth-only scenario (Scenario 3) or in the export-growth scenarios (Scenarios 1 and 2), and total food consumption per capita increases by 1.25–4.48 percent in the seven countries (Table A10, part C, column 7).

Scenario 5: Implications of Reduced Marketing Costs

Results of the first four scenarios suggest only limited impact of rapid growth on real GDP in each agricultural subsector, in spite of large multiplier effects in grain sector growth. In part, this is due to the small size of each of these subsectors (particularly nontraditional exports) relative to the overall economies. The scale of a sector is a factor in generating overall economic growth. For example, the combined four agricultural subsectors analyzed in Scenarios 1, 2, and 4 is around 10 percent of GDP in Mozambique, Zambia, and Zimbabwe, whereas growth in the three of them combined results in a GDP annual growth rate of 0.27–0.33 percent per capita in these three countries (with a fixed-labor-supply assumption, Table 4.2, part A, column 6). On the other hand, in Madagascar and Tanzania the combined four agricultural subsectors account for 24–25 percent of GDP, so growth in these four subsectors combined generates a per capita GDP annual growth of 0.82 percent and 0.76 percent, respectively (Table 4.2, part A, column 6).

---

[15]The current growth rate of livestock production in the region, as calculated from FAO data, is less than 3 percent.

This result suggests that investment and productivity gains outside the agricultural sector are needed to maximize growth linkages and overall benefits of agricultural growth. Spurring additional market demand is thus a key condition for enabling increases in agricultural production to lead to substantial increases in farmers' real incomes. Scenarios 5 and 6 explore such linkage effects.

Scenario 5 deals with the implications of reducing market costs in order to enhance effective demand for agricultural products. Although improving the performance of agricultural markets requires much more than investment in roads, communications, and other infrastructure,[16] in this scenario we focus on market efficiency by explicitly modeling a 4.5 percent increase in TFP in the transport sector, along with the productivity increases in the agricultural sector (combining Scenarios 1, 2, and 4). This increase in productivity reduces unit costs of transport and ultimately reduces input costs in other sectors of the economy. Moreover, the reduction in transport directly affects international trade by reducing the gap between domestic and border prices.[17] Improving the transportation sector's productivity lowers the unit cost of services provided by the sector, which reduces the market costs for trade in both agricultural and nonagricultural products, and thus increases marketing efficiency.

When growth in agriculture is combined with a more efficient transport sector, both total GDP and farm income increase sharply: additional GDP per capita annual growth rises to close to or more than 1 percent in five of the countries (all except Mozambique and Zimbabwe, Table 4.2, part A, column 7), and per capita farm income goes up by close to or more than 2 percent in all seven countries (Table A12, part B, column 2). Agricultural exports benefit more from improving the transport sector's productivity: total agricultural exports annually increase by 4.17–6.18 percent per capita in six countries and by 8.43 percent in Zambia (Table A12, part B, column 6). Malawi's agricultural export growth rate in this scenario is more than two times higher than the growth rate in the scenarios of agricultural growth alone. For the other six countries, agricultural export growth rates increase by 30 to 90 percent. Agricultural producer prices fall less than in the scenario with agricultural growth only, although the food consumption price index still falls in all the countries except Uganda. Total food consumption rises and livestock product consumption increases significantly (7–12 percent) in the seven countries, indicating the existence of potential market opportunities for agriculture, especially for livestock products.

Scenario 6: Easing the Domestic Demand Constraint by Combining Agricultural and Nonagricultural Growth

Another major potential source for increasing market demand for agricultural products is through growth in nonagricultural incomes. One avenue for achieving such growth is through the multiplier effects that emanate from agricultural growth (which were captured to some extent in the earlier simulations).[18] A second major avenue for increasing demand is through investments

---

[16]See, for example, the discussion of Poulton, Kydd, and Dorward (2006) about the factors and policies affecting market efficiency in African agriculture.

[17]These margins are calculated using the data of CIF and FOB prices for African countries with their trading partners bilaterally (Limao and Venables 2001).

[18]The CGE model used in this analysis, however, does not include the dynamic role of the regional and international market demand for stimulating growth, aspects that are better modeled in endogenous growth models (see Diao, Roe, and Yeldan 1999).

and increases in TFP in nonagricultural activities. Scenario 6 captures some of these interlinkages between agriculture and the rest of the economy by combining the exogenous productivity shocks included in Scenario 5 with an increase in TFP growth in some nonagricultural sectors of 3 percent per year. Specifically, a 3 percent TFP growth is assumed for textile and other labor-intensive manufacturing and private service sectors (including, apart from transportation, trade and other business-related private services). There is no additional productivity growth in the sectors dependent on natural resources and government services in the scenario. In total, the nonagricultural sectors (including the transport sector) modeled with high TFP growth account for more than 50 percent of national income in Malawi and Zambia, 40 percent in Mozambique and Zimbabwe, 30 percent in Tanzania and Madagascar, and 20 percent in Uganda.

The simulation results illustrate the importance of demand effects and the interlinkages between agriculture and nonagriculture. Improving growth performance of labor-intensive manufacturing and private services boosts total GDP in all seven countries (more than 2 percent of additional growth per capita annually in Malawi and Zambia, close to 2 percent in Madagascar and Tanzania, and 1.44 to 1.77 percent in Uganda, Zimbabwe, and Mozambique (Table 4.2, part A, column 8). Such growth in nonagriculture also spurs agricultural growth through increased demand for agricultural products. Farm income increases by 3.02 and 3.45 percent, respectively, in Uganda and Tanzania, and by close to or more than 4.0 percent per capita annually in the other five countries. Some 50–85 percent of the increase in farm income is largely achieved by nonagricultural productivity alone, and the smaller the contribution of agriculture to the economy (for example, in Zambia and Mozambique), the larger the share of the increased farm income that can be explained by nonagricultural productivity growth.

Per capita growth in total food consumption increases by 1.69 percent in Uganda, 2.80 percent in Mozambique, 3.21 and 3.83 percent in Tanzania and Zambia, 4.47 and 4.58 percent in Malawi and Madagascar, and 6.19 percent in Zimbabwe (Table A12, part C, column 3). Still, livestock consumption shows strong growth with high income elasticity (with 10–13 percent of annual growth in six countries and 7 percent, given a modest GDP growth, in Uganda). Although grain consumption would grow more slowly than income growth given that it is income inelastic, there is still 0.8–0.9 percent of growth per capita in grain consumption in the five countries and more than 1 percent in Uganda. In Madagascar, however, the growth rate is a very modest 0.2 percent annually. Moreover, in some of these countries growth in agricultural imports is more rapid than the growth in agricultural exports (for example, in Malawi and Zimbabwe agricultural imports grow at 8.5 and 5.5 percent, respectively), indicating that potential intraregional trade has to be further explored.

The 3 percent TFP annual growth in selected manufacturing and services may be overly optimistic in some countries given the investment, technological, and institutional constraints that inhibit growth in these countries.[19] Nonetheless, even though the mechanisms for achieving high productivity growth in nonagriculture are not specified, this scenario does highlight the demand-side role for agricultural growth

[19]Per capita GDP growth rates in sub-Saharan Africa averaged only 2.4 percent in 2003–04, however, implying a similar rate of TFP growth in all sectors (assuming labor force and capital growing at approximately the same growth rate as population). Nonetheless, a 3 percent TFP annual growth in overall GDP has been achieved in Ghana, Uganda, and other countries in Sub-Saharan Africa in recent years (World Bank World Development Report 2006).

and the importance of growth in nonagriculture for achieving growth in total income.

Summary

Results from the world CGE model of alternative growth scenarios for the seven sample African countries indicate that an export-led agricultural growth strategy without substantial reduction in marketing costs and an increase in nonagricultural productivity is unlikely to generate substantial overall income growth. Simulation of a recovery of their traditional exports to their historical high and regaining their lost markets results in an additional 0.08–0.15 percent annual growth in per capita GDP. Even with an assumption of more employment opportunities, the additional per capita GDP growth rate is only in the range of 0.16–0.47 percent per year. Given their relatively small size in most of the economies studied, simulated 10–24 percent of annual growth in nontraditional exports generates additional annual per capita GDP growth of only 0.01–0.05 percent. Even with an unemployment assumption, the greatest impact on per capita GDP growth among the seven countries is 0.11 percent a year (in Tanzania and Uganda).

The grain sector has stronger growth linkages with nonagriculture than the export-orientated agriculture sector does. However, without simultaneous growth in income generated outside the grain sector and significant substitution for imports through improving market channels, productivity in the grain sector can cause a shift in domestic terms of trade against agriculture, negating the income benefits of productivity improvement. Certain government interventions in markets to help stimulate import substitution can result in a significant decline in grain imports and a rise in farm income. Simultaneous productivity growth in food crops and livestock also offer more potential for increasing farm incomes.

However, accelerated productivity growth in all of these agricultural subsectors (traditional and nontraditional exports, grains, and livestock) has only a limited impact on real GDP owing to their relatively small size in each country's overall economy. Combined growth of these agricultural subsectors results in additional per capita GDP annual growth rate of 0.39–1.51 percent in the seven countries even with new employment opportunities.

Reduced marketing costs are crucial to reducing consumer food prices and raising producer incomes in the longer run. The model simulation combining productivity growth in agriculture with reduction in marketing costs suggests that the resulting increase in per capita GDP would be close to or more than 2 percent per year in Madagascar, Tanzania, and Uganda, with new employment opportunities. Another important way to accelerate growth in domestic food demand is by increasing incomes in the nonagricultural sectors. Some of the growth in nonagricultural productivity can be spurred by growth linkages, including those derived from additional spending of higher agricultural incomes by farm households. Agriculture can thus serve as one engine of growth, but economywide growth will likely require more than one engine. A second major avenue for increasing agricultural demand is through investments that increase productivity in the nonagricultural sector itself. With an optimistic assumption for growth in that sector, simulation results suggest a high payoff from the two-engine growth strategy. Combining similar agricultural and nonagricultural productivity growth results in much higher farm income growth, which reaches 3.0–4.4 percent per capita annually in the seven countries studied.

Three important caveats should be noted. First, the analysis presented here focuses explicitly on demand-side constraints of agricultural growth, assuming that subsector level growth is achievable. However, productivity growth itself is not a natural process and is constrained by many institutional, policy, and technological conditions. The underperformance of Africa's agriculture in the last four decades indicates that many components in these areas are not functioning effectively. Improving conditions involves not only a series of policy and institutional reforms, but also requires a significant increase in public and private investment. Many such reforms and investments will improve the overall environment of agricultural growth by increasing the sector's efficiency, productivity, and market competitiveness, including market development, institutional and policy reforms to encourage private-sector investment, public and private investment in infrastructure such as roads, telecommunications, and other information systems. However, certain sector-related investments are needed as well, and the cost of productivity growth can be significantly different across agricultural subsectors. For example, investment in nontraditional agricultural exports is often led by the private sector, whereas significant improvement of productivity in the staple crop and livestock sectors generally calls for increased public expenditure in agricultural research and extension and in market risk management systems. Owing to such differences in the costs of productivity enhancement among sectors, the calculated GDP growth effects driven by improvement in various agricultural subsectors should be interpreted with caution.

Second, the analysis has not assessed the technological feasibility of increased agricultural and nonagricultural productivity. Although some agricultural scientists and knowledgeable observers are optimistic about technological potential and some notable successes have been achieved,[20] it is by no means certain that suitable technological packages that are well adapted to diverse local conditions are available or that agricultural information services are effective in disseminating this technology. Moreover, other factors might also limit agricultural productivity growth, including severe labor constraints related to widespread HIV/AIDS in some areas, governance problems, and civil strife. Moreover, if rapid agricultural productivity growth should prove not to be feasible, the issue of demand constraints may not even arise.

Third, this analysis has not considered impacts of various alternative productivity shocks on income distribution and poverty. Though the simulations presented here provide information on the effects on labor earnings, agricultural incomes, and food consumption in aggregate, differences in effects for various types of households across countries are inevitable. Productivity increases in grains and other major food staples, in particular, are likely to have a greater impact on poverty and food consumption than growth in traditional and nontraditional export agriculture because of their broad base (see, for example, five countries' case studies in Diao et al. 2006).[21] Given that reduction in poverty and hunger is a major development objective, the analysis should be extended beyond consideration of impacts on economic growth and aggregate consumption.

---

[20]See, for example, Sanders, Shapiro, and Ramasamy 1996; Beyerlee and Eicher 1997; ICRISAT 2001; WARDA 2006. They argue that seed/fertilizer technologies that could significantly improve productivity of cereal crops are available for many medium- to high-potential zones of Africa.

[21]The five countries are Ethiopia, Ghana, Rwanda, Uganda, and Zambia.

CHAPTER 5

Conclusions

In today's more integrated world economy, agricultural growth in Sub-Saharan Africa crucially depends on improving African countries' competitiveness in the world agricultural market and thus expanding their marketing opportunities. This report focused on demand-side constraints and their implications for three broad alternative agricultural development strategies (promoting traditional export, nontraditional export, and food staple growth) in East and Southern Africa.

We have addressed three major questions. First, how constraining will demand be for future agricultural growth in Sub-Saharan Africa and is there sufficient demand for agricultural products to permit agriculture to grow at a rate that can significantly reduce poverty and hunger? Second, if technological change and increase in supply are achieved, which agricultural subsectors offer the greatest potential and can become more powerful engines for raising real incomes and increasing food consumption? Finally, what are the implications of reductions in marketing costs and growth linkages with nonagricultural sectors in achieving increasing market demand for agriculture? We addressed these issues through a combination of an analysis of historical trends in agricultural exports, a comparative advantage analysis in intraregional trade, and trade and income growth under various alternative scenarios in a world CGE simulation model.

Historical trends show that over the past two decades Sub-Saharan Africa has been suffering from declining terms of trade and substantial price instability for traditional exports. Besides limited growth in global demand for these traditional agricultural commodities and increasing competition from other countries, Sub-Saharan African countries are also constrained by low productivity, problems with maintaining quality, and asymmetric agricultural and trade policies of wealthy nations (especially production subsidies). Despite African countries' significant market reform initiatives, future growth prospects for the region's traditional agricultural exports do not appear promising unless their competitiveness is increased by improving productivity, product quality and variety, and marketing conditions.

Although Sub-Saharan Africa does have the potential to recover its lost world market share and volume of exports of traditional export crops, the CGE model simulations show that this would have only a small impact on real incomes for the countries studied. A major reason for the limited impact of an increase in traditional exports is that they account for a relatively small part of the GDP in these economies and have weak linkages with other domestic sectors. Unfortunately, world prices for these commodities are currently rather low, and a recovery in world prices to levels of the early 1970s is unlikely.

Our analysis suggests that demand need not constrain rapid growth in nontraditional exports. Given the low current levels of production and minute share of world trade, nontraditional export agriculture offers perhaps the most promising opportunities for growth. Nonetheless, to a large extent, nontraditional exports consist of numerous products that are targeted to niche markets. Some countries have achieved notable success (for example, Kenya's horticultural exports), but this success may not be so easily replicated, particularly where macroeconomic or political instability make private investment in export activities highly risky. The model simulation analysis for East and Southern Africa shows that even under optimistic assumptions about productivity growth in nontraditional export crops (for example, 10–24 percent annual export growth in the seven countries studied), acceleration of growth in this sector alone will have only a minimal effect on the growth of their real GDP.

Unlike most traditional and nontraditional agricultural exports, domestic demand and production of food crops and livestock products far exceeds internationally traded quantities in Africa. Supply of many staple commodities cannot meet with current demand, and total food demand in the region is projected to further increase considerably in the next 10–20 years (for example, cereal demand will be 60 percent higher than the current level by 2015). Existing and growing food demand in Africa offers a great potential for significant expansion of markets for many staple products. The regional comparative advantage analysis shows that many African countries import a long list of agricultural commodities from outside the region, including commodities that are exported by their regional neighbors, with food staples having the greatest growth potential for increased intraregional trade. However, the CGE model simulation analysis for East and Southern Africa shows that without simultaneous growth in income generated outside the sector, productivity growth in the grain sector can cause a shift in domestic terms of trade against agriculture, negating the income benefit of productivity improvement. Certain government interventions to help stimulate import substitution can help reduce these adverse terms-of-trade effects. Simultaneous productivity growth in food crops and livestock also offers potential for increasing farm incomes.

The CGE model simulations also indicate that simultaneous lowering of real marketing costs can significantly enhance the increases in farm incomes and food consumption from productivity growth in agriculture. Such a reduction in marketing costs would require investments in marketing infrastructure (roads and bridges, ports, storage facilities, electricity, and so on) and development of market institutions (see, for example, Kherallah et al. 2002). Supplementing productivity increases with a reduction in marketing costs results in an additional growth in per capita GDP to close to or more than 2 percent per year in the model simulations.

Accelerating growth in domestic demand for foods by increasing income in the nonagricultural sectors would also avoid negative terms-of-trade effects. Some of the growth in nonagricultural productivity can be spurred by growth linkages, including those deriving from additional farm household spending of higher agricultural incomes. Agriculture can thus serve as one engine of growth, but economywide growth will likely require more than one engine.[1] A second major avenue for increasing agricultural demand is through investments that increase productivity in the nonagricultural sector itself. Simulation results suggest that combined with growth in nonagricultural productivity, such growth in agriculture results in a much larger increase in agricultural real income, reaching 3.0–4.4 percent per capita annually in the seven countries studied.

A number of caveats are in order in interpreting these results, however. The structure of agriculture, the potential for agricultural exports, and many other economic and

[1]In fact, a structural transformation in which the nonagricultural sector grows faster than the agricultural sector is the typical pattern of economic growth. See Diao et al. (2006) for further analysis of such a structural transformation in Sub-Saharan Africa.

political factors vary enormously across countries in Africa. This report has focused mainly on the regional level, although the CGE model and simulations of alternative growth scenarios are for the seven East and Southern African countries with available data. Obviously, an individual country's circumstances, such as its stage of development, degree of urbanization, geography and size, and, more importantly, the developmental capability of its governments, all affect the types of state interventions and policy choices. Given the existing diversities across countries and regions in Sub-Saharan Africa, further work is needed for both development of typologies and for country-level assessments of alternative agricultural development strategies. Moreover, differences in effects for various types of households are inevitable across countries. Productivity increases in grains and other major food staples, in particular, are likely to have larger effects on poverty and food consumption than growth in traditional and nontraditional export agriculture in the early stages of development because of the broad base. Given that reduction in poverty and hunger is a major development objective, the analysis should be extended beyond consideration of impacts on economic growth and aggregate consumption.

Nonetheless, the results presented here have clear implications for the design of agricultural development and long-term food security strategies. First, overall demand constraints need not limit agricultural productivity growth for nontraditional exports in the medium term. There is also some scope for increased effective demand for food grains, especially in conjunction with programs that expand the livestock sector. Trade restrictions and domestic pricing policies are particularly important, and better integration of regional markets could provide important outlets for increased grain production in some countries.

Second, there is a high payoff to reducing transaction costs in agricultural markets. In the short run, rapid increases in agricultural production can lead to sharp price declines where markets are thin and underdeveloped. Reduced marketing costs are crucial to reducing consumer food prices, and in the longer run raise producer incomes as well. Operationally, public investment in marketing infrastructure (roads, electricity, telecommunications) involve tough choices and will require rigorous cost-benefit analyses to assess priorities for investment across regions (for example, near urban centers and ports to promote international trade versus roads in more isolated areas) and the types of investments (for example, building feeder roads versus main roads). Policy frameworks that encourage private investment in cold storage and agricultural processing activities are also needed. Thus, well-designed investments in infrastructure and policy reforms that lead to reductions in transaction costs have the potential to benefit a wide spectrum of agricultural (and nonagricultural) activities and avoid the problem of requiring the government to pick "winners" among competing crops or agricultural activities.

Finally, without growth in the nonagricultural sector, gains in overall incomes and calorie consumption for Sub-Saharan Africa as a whole will be limited. Thus, investments in agriculture and other efforts to promote greater agricultural productivity have to be in tandem with policies and investments to spur nonagricultural growth. Moreover, investments in rural infrastructure can help to maximize positive linkage effects of agricultural growth. Agricultural growth can play a major role in increasing food production, but sustained increases in incomes and reductions in poverty are likely to require a combination of labor-intensive growth in both agricultural and nonagricultural activities.

APPENDIX A

Tables

Table A1 Major coffee exporting countries in Sub-Saharan Africa

	Average annual exports (million tons)		Ranking in Sub-Saharan Africa (order)	
	1970–74	2000–03	1970–74	2000–03
Angola	195,037	1,157	3	20
Benin	1,679	4	22	25
Burundi	21,578	20,275	11	7
Cameroon	75,021	62,878	4	4
Central African Republic	8,935	5,028	13	13
Congo, Democratic Republic of	69,080	11,430	7	9
Congo, Republic of	860	3,127	21	17
Côte d'Ivoire	208,922	185,650	1	1
Equatorial Guinea	5,784	322	18	22
Ethiopia	73,184	117,624	6	3
Gabon	379	40	24	24
Ghana	4,430	2,096	17	18
Guinea	7,492	11,021	16	10
Kenya	64,129	55,527	5	5
Liberia	5,288	334	15	21
Madagascar	58,165	4,815	8	14
Malawi	173	3,421	25	16
Mali	988	0	23	26
Nigeria	2,674	305	20	23
Rwanda	16,989	17,228	10	8
Sierra Leone	8,019	1,896	14	19
Tanzania	47,310	44,710	9	6
Togo	11,592	6,236	12	11
Uganda	201,078	129,315	2	2
Zambia	0	5,270	26	12
Zimbabwe	1,130	4,109	19	15
Subtotal	1,089,915	693,818		
Total Sub-Saharan Africa	1,090,094	693,908		

Source: FAOSTAT (2006), last accessed in March 2006.

Table A2 Sub-Saharan African major agricultural export commodities, 1996–2000 annual average

		SSA exports to the world		SSA exports excluding SACU to the world		SSA intraregional exports		SSA intraregional exports, excluding SACU		
	Rank[a]	Value (million US$)	Share (%)	Value (million US$)	Share (%)	Value (million US$)	Share (%)	Value (million US$)	Share (%)	SSA intraregional exports/ total SSA exports
Staples										
Meat	16	335	1.8	120	0.9	94	5.0	33	2.9	28.1
Livestock	18	123	0.7	115	0.8	19	1.0	13	1.1	15.1
Other cereals	19	207	1.1	104	0.7	160	8.5	82	7.2	77.2
Maize	22	287	1.6	66	0.5	150	8.0	55	4.9	52.1
Beans	24	54	0.3	48	0.3	18	1.0	14	1.2	33.6
Cassava	28	2	0.0	2	0.0	0	0.0	0	0.0	2.6
Oilseeds	12	317	1.7	271	1.9	30	1.6	23	2.0	9.6
Oils and fat	13	359	2.0	270	1.9	140	7.5	89	7.8	39.1
Sugar	8	1,063	5.8	605	4.3	159	8.5	76	6.7	14.9
Total staples		2,747	14.9	1,600	11.5	769	41.1	385	33.8	28.0
Nontraditional										
Fish	2	2,403	13.1	1,841	13.2	209	11.2	171	15.0	8.7
Vegetables and fruits	6	2,449	13.3	976	7.0	135	7.2	38	3.4	5.5
Miscellaneous	7	777	4.2	610	4.4	83	4.4	71	6.2	10.7
Processed cocoa	10	492	2.7	470	3.4	21	1.1	12	1.0	4.3
Coffee roasted	21	105	0.6	100	0.7	29	1.5	24	2.1	27.4
Processed food	20	159	0.9	100	0.7	110	5.9	63	5.5	69.1
Beverages	27	334	1.8	27	0.2	82	4.4	20	1.7	24.5
Total nontraditional		6,719	36.5	4,124	29.6	669	35.8	399	35.0	10.0

Traditional										
Cocoa bean	1	2,387	13.0	2,386	17.1	9	0.5	9	0.8	0.4
Coffee green	3	1,844	10.0	1,831	13.2	44	2.4	43	3.8	2.4
Cotton	4	1,459	7.9	1,421	10.2	102	5.5	101	8.9	7.0
Tobacco	5	1,125	6.1	1,060	7.6	63	3.4	57	5.0	5.6
Tea	9	646	3.5	592	4.3	60	3.2	55	4.8	9.2
Cashew nuts	14	221	1.2	220	1.6	2	0.1	2	0.2	0.8
Nuts	25	44	0.2	31	0.2	6	0.3	5	0.5	13.4
Other fibers	26	233	1.3	31	0.2	5	0.3	1	0.1	2.2
Total traditional		7,959	43.3	7,571	54.4	291	15.6	274	24.0	3.7
Other										
Animal skins	11	564	3.1	311	2.2	8	0.4	6	0.5	1.4
Spices	15	157	0.9	139	1.0	8	0.5	4	0.4	5.4
Feedstuff	17	147	0.8	118	0.8	41	2.2	27	2.3	27.8
Cigarettes	23	108	0.6	60	0.4	82	4.4	44	3.8	76.5
Total other		975	5.3	627	4.5	140	7.5	81	7.1	14.3
Total		18,400	100	13,923	100	1,870	100	1,139	100	10.2

Source: UNCOMTRADE (2005), last accessed in April 2005.

Note: This table does not include wood products, rubber, and textile exports, which are included in the United Nations definition of agricultural imports. SSA is Sub-Saharan Africa; SACU is Southern African Customs Union.

[a]By share in Sub-Saharan Africa's total agricultural exports, excluding exports from SACU.

Table A3 East African major agricultural export commodities, 1996–2000 annual average

	Rank[a]	East African exports to the world: Value (million US$)	East African exports to the world: Share (%)	East African exports to SSA: Value (million US$)	East African exports to SSA: Share (%)	East African intraregional exports: Value (million US$)	East African intraregional exports: Share (%)	East African intraregional exports/Total East African exports (%)
Staples								
Livestock	9	111	2.5	2	0.4	1	0.3	1.4
Other cereals	15	50	1.1	28	7.9	26	8.8	55.7
Beans	16	44	1.0	13	3.6	11	3.6	28.9
Meat	17	38	0.8	4	1.1	3	1.1	10.0
Maize	18	30	0.7	27	7.7	19	6.4	90.0
Cassava	28	0	0.0	0	0.0	0	0.0	7.0
Oilseeds	7	162	3.6	3	0.9	3	0.9	1.9
Oils and fat	12	87	1.9	34	9.5	33	11.1	39.1
Sugar	14	80	1.8	17	4.9	17	5.7	21.7
Total staples		603	13.4	128	35.9	114	38.0	21.2
Nontraditional								
Fish	3	474	10.5	13	3.6	6	2.1	2.7
Miscellaneous	4	342	7.6	49	13.8	48	15.9	14.4
Vegetables and fruits	5	330	7.3	15	4.3	13	4.3	4.6
Processed food	22	19	0.4	17	4.9	17	5.7	90.3
Beverages	24	11	0.2	8	2.3	8	2.6	77.3
Coffee roasted	23	14	0.3	3	0.9	3	1.1	23.5
Processed cocoa	26	2	0.0	1	0.4	1	0.4	78.8
Total nontraditional		1,192	26.5	107	30.1	96	32.1	9.0

Traditional								
Coffee green	1	1,406	31.3	35	9.7	25	8.2	2.5
Tea	2	524	11.6	38	10.7	34	11.2	7.3
Cotton	6	226	5.0	11	3.2	7	2.4	5.0
Tobacco	11	100	2.2	10	2.7	3	1.0	9.5
Cashew nuts	13	81	1.8	0	0.0	0	0.0	0.2
Other fibers	19	30	0.7	1	0.3	0	0.0	3.9
Cocoa bean	24	13	0.3	0	0.0	0	0.0	0.1
Nuts	27	7	0.2	0	0.0	0	0.0	1.2
Total traditional		2,387	53.1	95	26.6	68	22.8	4.0
Other								
Animal skin	8	159	3.5	3	0.7	1	0.3	1.6
Spices	10	110	2.5	2	0.4	1	0.3	1.4
Feedstuff	20	24	0.5	4	1.0	1	0.4	15.3
Cigarettes	21	20	0.5	18	5.1	18	5.9	89.9
Total other		314	7.0	26	7.4	21	7.0	8.3
Total		4,496	100	357	100	300	100	7.9

Source: UNCOMTRADE (2005), last accessed in April 2005.
Note: This table does not include wood products, rubber, and textile exports, which are included in the United Nations definition of agricultural imports. SSA is Sub-Saharan Africa.
[a]By share in East African total agricultural exports.

Table A4 Southern African major agricultural export commodities, 1996–2000 annual average

		Southern African exports to the world		Southern African exports, excluding SACU, to the world		Southern African exports to SSA		Southern African exports, excluding SACU, to SSA		Southern African intraregional exports		Southern African intraregional exports/ total Southern African exports (%)
	Rank[a]	Value (million US$)	Share (%)	Value (million US$)	Share (%)	Value (million US$)	Share (%)	Value (million US$)	Share (%)	Value (million US$)	Share (%)	
Staples												
Meat	7	287	4.9	65	3.3	80	8.8	18	6.2	62	8.9	28.0
Maize	10	258	4.4	37	1.9	120	13.1	30	10.4	70	10.0	46.5
Other cereals	11	143	2.4	37	1.9	112	12.2	34	11.5	76	11.0	78.0
Livestock	19	16	0.3	8	0.4	11	1.2	6	1.9	11	1.5	70.6
Beans	21	13	0.2	6	0.3	8	0.8	2	0.8	7	1.0	58.5
Cassava	27	0	0.0	0	0.0	0	0.0	0	0.0	0	0.0	1.9
Oilseeds	15	67	1.1	23	1.2	21	2.3	14	4.9	19	2.8	31.0
Oils and fat	20	95	1.6	7	0.4	57	6.2	6	2.1	44	6.3	60.3
Sugar	2	933	15.9	491	25.0	121	13.2	46	15.7	70	10.1	12.9
Total staples		879	15.0	183	9.3	409	44.7	110	37.9	289	41.6	46.5
Nontraditional												
Fish	3	777	13.2	228	11.6	48	5.2	11	3.8	23	3.3	6.2
Miscellaneous	5	266	4.5	96	4.9	16	1.7	4	1.4	12	1.8	6.0
Vegetables and fruits	6	1,520	25.9	78	4.0	100	10.9	10	3.4	82	11.8	6.6
Processed food	18	74	1.3	10	0.5	54	5.9	9	3.1	40	5.8	73.6
Beverages	22	298	5.1	7	0.3	70	7.7	5	1.8	52	7.4	23.5
Coffee roasted	24	6	0.1	1	0.1	5	0.6	1	0.2	4	0.6	84.8
Processed cocoa	26	22	0.4	0	0.0	10	1.1	0	0.1	8	1.1	44.6
Total nontraditional		2,935	50.0	419	21	288	31.5	40	14	209	30.1	9.8

Traditional												
Tobacco	1	973	16.6	943	48.0	54	5.9	45	15.3	46	6.7	5.5
Cotton	4	211	3.6	186	9.5	48	5.3	47	16.3	47	6.7	22.9
Tea	8	99	1.7	51	2.6	15	1.7	11	3.8	13	1.9	15.3
Coffee green	9	63	1.1	49	2.5	5	0.5	3	1.2	4	0.6	7.2
Cashew nuts	12	35	0.6	35	1.8	1	0.1	1	0.3	1	0.2	3.4
Nuts	23	14	0.2	3	0.1	1	0.1	0	0.1	1	0.1	5.4
Other fibers	25	217	3.7	0	0.0	4	0.4	0	0.0	4	0.5	1.8
Cocoa bean	28	1	0.0	0	0.0	0	0.0	0	0.0	0	0.0	0.8
Total traditional		1,612	27.5	1,267	64.5	127	13.9	108	37.2	116	16.8	7.9
Other												
Animal skins	13	286	4.9	29	1.5	6	0.6	3	1.0	6	0.8	2.0
Feedstuff	14	58	1.0	25	1.3	32	3.5	16	5.7	30	4.3	55.1
Spices	16	39	0.7	22	1.1	6	0.7	2	0.7	6	0.8	15.7
Cigarettes	17	62	1.1	18	0.9	47	5.1	12	4.0	39	5.6	75.7
Total other		445	7.6	94	4.8	91	9.9	33	11.4	80	11.5	20.4
Total		5,870	100	1,964	100	915	100	291	100	695	100	15.6

Source: UNCOMTRADE (2005), last accessed in April 2005.

Note: This table does not include wood products, rubber, and textile exports, which are included in the United Nations definition of agricultural imports. SSA is Sub-Saharan Africa; SACU is Southern African Customs Union.

[a]By share in Southern African (excluding SACU) total agricultural exports. Data for individual countries of SACU were not available. We use SACU to represent South Africa.

Table A5 Sub-Saharan African major agricultural import commodities, 1996–2000 annual average

		SSA imports from the world		SSA imports, excluding SACU, from the world		SSA intraregional imports		SSA intraregional imports, excluding SACU		SSA intraregional imports/total SSA imports
	Rank[a]	Value (million US$)	Share (%)	Value (million US$)	Share (%)	Value (million US$)	Share (%)	Value (million US$)	Share (%)	(%)
Staples										
Other cereals	1	2,572	25.4	2,218	27.7	160	8.5	135	8.7	6.2
Meat	3	1,104	10.9	894	11.2	94	5.0	74	4.8	8.5
Maize	11	272	2.7	226	2.8	150	8.0	138	9.0	54.9
Beans	17	105	1.0	80	1.0	18	1.0	16	1.0	17.1
Livestock	24	35	0.3	27	0.3	19	1.0	16	1.1	53.2
Cassava	28	0	0.0	0	0.0	0	0.0	0	0.0	39.7
Oils and fat	2	1,239	12.2	901	11.3	140	7.5	132	8.6	11.3
Oilseeds	21	56	0.6	33	0.4	30	1.6	20	1.3	54.3
Sugar	4	797	7.9	756	9.5	159	8.5	135	8.8	19.9
Total staples		6,181	61.0	5,134	64.2	769	41.1	667	43.3	12.4
Nontraditional										
Fish	5	787	7.8	709	8.9	209	11.2	203	13.2	26.6
Processed food	6	554	5.5	480	6.0	110	5.9	99	6.4	19.8
Beverages	7	456	4.5	315	3.9	82	4.4	75	4.9	18.0
Vegetables and fruits	8	435	4.3	346	4.3	135	7.2	109	7.1	31.1
Miscellaneous	9	333	3.3	182	2.3	83	4.4	65	4.2	25.0
Coffee roasted	18	76	0.7	69	0.9	29	1.5	28	1.8	37.9
Processed cocoa	20	60	0.6	32	0.4	21	1.1	12	0.8	35.7
Total nontraditional		2,699	26.7	2,132	26.7	669	35.8	591	38.4	24.8

Traditional										
Tobacco		148	1.5	86	1.1	63	3.4	27	1.7	42.9
Tea	14	130	1.3	111	1.4	60	3.2	46	3.0	46.0
Cotton	16	119	1.2	63	0.8	102	5.5	51	3.3	86.0
Coffee green	19	73	0.7	41	0.5	44	2.4	38	2.5	60.4
Other fibers	22	44	0.4	27	0.3	5	0.3	4	0.3	11.7
Nuts	25	22	0.2	8	0.1	6	0.3	5	0.3	27.0
Cocoa bean	26	12	0.1	5	0.1	9	0.5	2	0.1	70.2
Cashew nuts	27	4	0.0	1	0.0	2	0.1	0	0.0	47.1
Total traditional		552	5.5	341	4.3	291	15.6	173	11.3	52.8
Other										
Cigarettes	10	305	3.0	280	3.5	82	4.4	76	5.0	27.1
Feedstuff	12	224	2.2	66	0.8	41	2.2	21	1.4	18.2
Animal skins	15	124	1.2	20	0.2	8	0.4	4	0.3	6.6
Spices	23	41	0.4	21	0.3	8	0.5	5	0.4	20.6
Total other		693	6.8	387	4.8	140	7.5	107	7.0	20.2
Total		10,125	100	7,994	100	1,870	100	1,539	100	18.5

Source: UNCOMTRADE (2005), last accessed in April 2005.

Note: This table does not include wood products, rubber, and textile exports, which are included in the United Nations definition of agricultural imports. SSA is Sub-Saharan Africa; SACU is Southern African Customs Union.

[a]By share in Sub-Saharan Africa's total agricultural imports, excluding imports from SACU.

Table A6 East African major agricultural import commodities, 1996–2000 annual average

		East African imports from the world		East African imports from SSA		East African intraregional imports		East African intraregional imports/ total East African imports (%)
	Rank[a]	Value (million US$)	Share (%)	Value (million US$)	Share (%)	Value (million US$)	Share (%)	
Staples								
Other cereals	1	621	29.0	56	11.0	26	8.8	9.0
Meat	4	148	6.9	17	3.2	3	1.1	11.1
Maize	5	113	5.3	67	13.0	19	6.4	59.1
Beans	12	53	2.5	11	2.2	11	3.6	21.0
Livestock	23	5	0.2	1	0.3	1	0.3	30.0
Cassava	28	0	0.0	0	0.0	0	0.0	99.4
Oils and fat	2	406	18.9	46	8.9	33	11.1	11.2
Oilseeds	19	8	0.4	4	0.8	3	0.9	50.2
Sugar	3	185	8.6	60	11.7	17	5.7	32.4
Total staples		1,539	71.8	261	51.2	114	38.0	17.0
Nontraditional								
Processed food	6	112	5.2	28	5.5	17	5.7	25.3
Miscellaneous	7	81	3.8	50	9.8	48	15.9	61.5
Vegetables and fruits	8	76	3.5	24	4.7	13	4.3	31.7
Fish	9	70	3.2	12	2.4	6	2.1	17.4
Beverages	10	68	3.2	23	4.6	8	2.6	34.6
Processed cocoa	18	10	0.5	3	0.6	1	0.4	29.8
Coffee roasted	20	7	0.3	4	0.7	3	1.1	57.4
Total nontraditional		422	19.7	144	28.3	96	32.1	34.2

Traditional								
Tea	13	42	1.9	35	6.8	34	11.2	83.2
Coffee green	14	27	1.3	25	5.0	25	8.2	92.3
Cotton	17	10	0.5	9	1.8	7	2.4	86.3
Tobacco	16	14	0.7	8	1.5	3	1.0	54.6
Other fibers	21	5	0.2	0	0.0	0	0.0	4.5
Nuts	25	1	0.0	0	0.0	0	0.0	12.2
Cocoa bean	26	0	0.0	0	0.0	0	0.0	12.3
Cashew nuts	27	0	0.0	0	0.0	0	0.0	19.2
Total traditional		100	4.7	77	15.1	68	22.8	77.0
Other								
Cigarettes	11	57	2.6	23	4.4	18	5.9	39.9
Feedstuff	15	17	0.8	3	0.5	1	0.4	16.6
Spices	22	5	0.2	1	0.3	1	0.3	26.0
Animal skins	24	3	0.1	1	0.2	1	0.3	40.1
Total other		81	3.8	28	5.5	21	7.0	34.3
Total		2,143	100	511	100	300	100	23.8

Source: UNCOMTRADE (2005), last accessed in April 2005.

Note: This table does not include wood products, rubber, and textile exports, which are included in the United Nations definition of agricultural imports. SSA is Sub-Saharan Africa.

[a]By share in East African total agricultural imports.

Table A7 Southern African major agricultural import commodities, 1996–2000 annual average

		Southern African imports from the world		Southern African imports, excluding SACU, from the world		Southern African imports from SSA		Southern African imports, excluding SACU, from SSA		Southern African intraregional imports		Southern African intra-regional imports/total Southern African imports (%)
	Rank[a]	Value (million US$)	Share (%)	Value (million US$)	Share (%)	Value (million US$)	Share (%)	Value (million US$)	Share (%)	Value (million US$)	Share (%)	
Staples												
Meat	2	426	14.8	206	18.3	62	8.5	43	9.6	62	9.7	14.5
Maize	6	156	5.4	90	8.0	78	10.7	61	13.7	70	11.1	49.9
Other cereals	1	645	22.4	274	24.4	77	10.6	52	11.6	76	12.1	12.0
Livestock	18	20	0.7	11	1.0	11	1.6	9	2.0	11	1.7	57.4
Beans	15	53	1.9	22	2.0	9	1.2	5	1.2	7	1.1	16.8
Cassava	28	0	0.0	0	0.0	0	0.0	0	0.0	0	0.0	11.7
Oilseeds	17	45	1.6	13	1.1	21	2.9	11	2.4	19	3.1	46.3
Oils and fat	3	508	17.7	152	13.5	44	6.1	37	8.2	44	7.0	8.8
Sugar	7	124	4.3	89	7.9	70	9.7	53	11.8	70	11.1	57.0
Total staples		1,301	45.3	603	53.6	237	32.7	171	38.0	226	35.7	18.3
Nontraditional												
Fish	9	143	5.0	64	5.7	28	3.8	21	4.7	23	3.6	19.3
Miscellaneous	12	178	6.2	25	2.2	27	3.8	9	2.1	12	2.0	15.4
Vegetables and fruits	5	183	6.4	103	9.1	84	11.6	66	14.7	82	13.0	46.3
Processed food	8	144	5.0	71	6.3	41	5.6	32	7.0	40	6.4	28.3
Beverages	4	246	8.6	106	9.4	52	7.2	43	9.6	52	8.2	21.1
Coffee roasted	22	12	0.4	6	0.5	5	0.6	4	0.8	4	0.7	36.4
Processed cocoa	19	37	1.3	11	0.9	16	2.2	7	1.5	8	1.3	42.3
Total nontraditional		893	31.1	368	32.7	232	31.9	171	38.2	209	33.2	26.0

Traditional												
Tobacco	14	84	2.9	23	2.1	52	7.2	18	3.9	46	7.3	62.3
Cotton	11	90	3.1	26	2.3	77	10.6	23	5.0	47	7.4	86.0
Tea	23	22	0.8	4	0.3	16	2.3	3	0.8	13	2.1	74.2
Coffee green	24	39	1.3	4	0.3	10	1.3	3	0.7	4	0.7	25.0
Cashew nuts	26	3	0.1	0	0.0	1	0.2	0	0.0	1	0.2	48.9
Nuts	25	15	0.5	2	0.2	1	0.2	0	0.1	1	0.1	8.4
Other fibers	16	32	1.1	14	1.3	5	0.6	4	0.8	4	0.6	14.2
Cocoa bean	27	8	0.3	0	0.0	7	1.0	0	0.0	0	0.0	97.6
Total traditional		293	10.2	73	6.5	171	23.5	51	11.4	116	18.4	58.2
Other												
Animal skins	20	116	4.0	9	0.8	7	1.0	3	0.6	6	0.9	6.2
Feedstuff	13	184	6.4	24	2.2	34	4.6	13	2.8	30	4.7	18.2
Spices	21	25	0.9	6	0.5	6	0.9	4	0.8	6	0.9	25.1
Cigarettes	10	61	2.1	41	3.6	40	5.5	37	8.2	39	6.2	65.1
Total other		387	13.5	80	7.1	87	11.9	56	12.4	80	12.7	22.5
Total		2,873	100	1,124	100	727	100	449	100	632	100	25.3

Source: UNCOMTRADE (2005), last accessed in April 2005.
Note: This table does not include wood products, rubber, and textile exports, which are included in the United Nations definition of agricultural imports. SSA is Sub-Saharan Africa; SACU is Southern African Customs Union.
[a]By share in Southern African (excluding SACU) total agricultural imports. Data for individual countries of SACU were not available. We use SACU to represent South Africa.

Table A8 Economic structure of seven low-income countries in East and Southern Africa

	Malawi	Madagascar	Mozam-bique	Tanzania	Uganda	Zambia	Zimbabwe
A. GDP and AgGDP							
GDP (current million US$)	1,714	5,474	4,321	10,297	6,297	4,335	17,750
Population (million)	11.0	16.9	18.8	35.9	25.3	10.4	13.1
GDP per capita (constant 2000 US$)	157	233	255	309	277	354	479
Agricultural GDP (current million US$)	601	1,465	1,013	4,254	1,867	897	2,737
AgGDP share of GDP (%)	38.4	29.2	26.1	45.0	32.4	22.8	17.4
Exports/GDP (%)	25.4	16.0	23.5	17.1	11.9	23.7	24.1
Imports/GDP (%)	43.7	22.6	38.2	24.6	26.5	29.0	21.8
B. Share of AgGDP in 2002 (%)							
Food crops	33.9	27.4	25.5	40.8	14.9	27.9	19.9
Cereals	28.1	24.3	23.2	33.2	12.2	22.2	12.4
Maize	27.5	0.7	19.7	26.5	11.1	19.1	9.7
Other crops	55.7	38.3	59.4	44.9	67.8	57.9	59.8
Cotton	1.4	1.1	2.2	1.1	0.9	4.4	10.6
Exportable fruits and vegetables	8.0	5.2	23.4	10.5	55.7	7.0	0.6
Exportable tree crops	3.2	9.5	25.0	18.0	0.9	29.3	9.1
Livestock	10.4	34.3	15.1	14.2	17.3	14.2	20.4

Source: World Bank (2006), last accessed in March 2006.
Note: Data for GDP and AgGDP are for 2003, except for Zimbabwe data, which are for 2002.

Table A9 Economywide effects of increase in traditional exports in Simulation 1 (additional per capita annual growth rate, %)

	Traditional export sector					Food consumption and prices		Agricultural trade		Returns to land and labor		
	Production	Exports	Producer prices	Real GDP	Agri-cultural GDP	Con-sumption	Price index	Imports	Exports	Land	Unskilled labor	Skilled labor
A. With fixed labor supply												
Malawi	1.8	1.9	–0.10	0.09	–0.39	0.25	–0.16	4.14	1.37	–1.44	0.14	0.15
Madagascar	5.1	10.3	0.32	0.14	0.62	0.24	–0.01	1.72	2.83	0.79	0.34	0.07
Mozambique	9.1	15.0	0.21	0.08	0.56	0.16	0.01	0.95	2.89	0.72	0.19	0.05
Tanzania	5.7	10.1	0.26	0.15	0.59	0.26	0.09	2.29	2.87	0.67	0.34	0.04
Uganda	10.0	10.2	0.44	0.14	0.64	0.09	0.43	3.46	4.71	1.00	0.52	0.09
Zambia	5.0	10.3	0.20	0.10	0.52	0.19	0.00	–0.17	3.89	0.62	0.23	0.08
Zimbabwe	3.2	3.6	0.06	0.11	–0.40	0.24	–0.10	2.19	2.38	–0.50	0.13	0.19
B. With unemployed unskilled labor												
Malawi	1.9	2.0	–0.11	0.16	–0.33	0.34	–0.18	4.19	1.45	–1.36		0.21
Madagascar	5.6	11.2	0.16	0.34	0.78	0.46	–0.09	1.47	3.37	1.01		0.28
Mozambique	9.3	15.2	0.15	0.17	0.70	0.29	–0.02	0.89	3.10	0.90		0.11
Tanzania	6.1	10.6	0.16	0.33	0.75	0.50	0.02	2.18	3.28	0.92		0.24
Uganda	11.0	11.2	0.26	0.47	0.96	0.57	0.23	3.31	5.53	1.49		0.23
Zambia	5.2	10.7	0.14	0.21	0.67	0.34	–0.03	–0.28	4.27	0.84		0.15
Zimbabwe	3.3	3.7	0.05	0.16	–0.37	0.30	–0.11	2.20	2.47	–0.46		0.23

Sources: CGE model results.

Table A10 Economywide effects of increase in grain and livestock production in Simulations 3, 3a, and 4 (with fixed labor supply) (additional per capita annual growth rate, %)

	Grain sector						Food consumption and prices		Agricultural trade		Returns to land and labor		
	Grain production	Grain consumption	Grain imports	Producer prices[a]	Real GDP	Agricultural GDP	Total consumption	Price index	Imports	Exports	Land	Unskilled Labor	Skilled Labor
A. Growth in grain sector only													
Malawi	1.41	0.43	–8.78	–2.10	0.19	–0.31	0.40	–0.72	–1.74	–0.18	–0.51	0.05	0.18
Madagascar	0.74	0.35	–7.37	–2.08	0.18	–0.12	0.29	–0.46	–1.39	0.14	–0.33	0.06	0.29
Mozambique	2.13	0.61	–4.24	–2.29	0.11	0.03	0.31	–0.38	–0.54	0.25	0.01	0.09	0.08
Tanzania	1.41	0.50	–4.57	–2.03	0.27	–0.19	0.43	–0.44	–1.74	0.49	–0.35	0.09	0.44
Uganda	1.90	1.06	–4.45	–2.33	0.15	0.10	0.27	–0.18	–1.32	–0.10	0.06	0.11	0.13
Zambia	0.77	0.37	–2.45	–1.46	0.09	–0.20	0.25	–0.37	–0.77	0.33	–0.29	0.02	0.10
Zimbabwe	0.55	0.23	–5.45	–1.37	0.03	–0.34	0.11	–0.33	–0.78	0.06	–0.33	–0.02	0.04
B. With grain trade facility													
Malawi	3.60	0.48	–51.78	–1.84	0.23	0.60	0.05	–0.90	–8.95	–0.99	0.62	0.29	0.03
Madagascar	1.81	0.32	–35.50	–2.05	0.18	–0.14	0.15	–0.52	–6.69	–0.10	–0.30	0.11	0.26
Mozambique	12.12	0.65	–65.42	–1.47	0.20	1.65	0.03	0.13	–8.20	2.43	2.16	1.01	0.37
Tanzania	4.21	0.31	–27.19	–2.10	0.26	–0.49	–0.17	–0.62	–9.77	–0.20	–0.58	–0.02	0.49
Uganda	6.30	1.21	–41.55	–2.25	0.18	0.09	–0.02	–0.23	–12.82	–2.56	0.01	0.23	0.15
Zambia	3.28	0.30	–38.14	–1.34	0.11	0.00	–0.12	–0.45	–9.35	–0.14	–0.08	0.18	0.19
Zimbabwe	1.16	0.20	–22.47	–1.37	0.04	–0.27	0.01	–0.37	–3.00	–0.10	–0.33	–0.02	0.02
C. With growth in livestock													
Malawi	1.12	0.11	–8.29	–1.99	0.35	1.16	2.57	–0.61	–0.81	–0.12	–0.44	0.26	0.12
Madagascar	0.21	–0.17	–5.94	–1.60	0.64	1.27	3.59	–0.73	–0.74	–0.65	0.94	0.70	0.22
Mozambique	1.86	0.39	–3.94	–2.14	0.22	0.63	1.39	–0.46	–0.41	0.39	0.46	0.24	0.09
Tanzania	1.11	0.28	–4.49	–1.91	0.54	0.37	1.88	–0.70	–1.51	0.58	–0.37	0.32	0.78
Uganda	1.59	0.75	–4.16	–2.03	0.49	0.71	1.25	–0.34	–1.29	–0.33	0.49	0.48	0.09
Zambia	0.72	0.02	–3.01	–1.40	0.19	0.18	2.00	–0.54	–0.89	0.44	–0.25	0.15	0.19
Zimbabwe	0.60	–0.02	–5.39	–1.19	0.16	1.27	4.48	–0.54	–0.61	–0.15	0.92	0.18	–0.07

Source: CGE model results.

[a]For maize and other coarse grains.

Table A11 Economywide effects of increase in grain and livestock production in Simulations 3, 3a, and 4 (with unemployed unskilled labor) (additional per capita annual growth rate, %)

	Grain sector						Food consumption/prices		Agricultural trade		Returns to land and labor	
	Grain production	Grain consumption	Grain imports	Producer prices[a]	Real GDP	Agricultural GDP	Total consumption	Food price index	Imports	Exports	Land	Skilled labor
A. Growth in grain sector only												
Malawi	1.44	0.45	–8.87	–2.12	0.22	–0.28	0.43	–0.73	–1.73	–0.15	–0.48	0.20
Madagascar	0.77	0.37	–7.47	–2.11	0.21	–0.09	0.33	–0.47	–1.43	0.24	–0.29	0.33
Mozambique	2.19	0.64	–4.28	–2.32	0.16	0.10	0.37	–0.40	–0.57	0.37	0.10	0.11
Tanzania	1.46	0.53	–4.63	–2.06	0.32	–0.15	0.49	–0.46	–1.77	0.61	–0.29	0.49
Uganda	2.00	1.14	–4.53	–2.39	0.23	0.17	0.37	–0.22	–1.37	0.10	0.17	0.16
Zambia	0.77	0.38	–2.36	–1.46	0.10	–0.19	0.26	–0.37	–0.78	0.38	–0.27	0.10
Zimbabwe	0.54	0.23	–5.43	–1.37	0.03	–0.34	0.10	–0.33	–0.77	0.04	–0.34	0.04
B. With grain trade facility												
Malawi	3.76	0.56	–52.22	–1.97	0.39	0.73	0.25	–0.97	–8.69	–0.79	0.74	0.16
Madagascar	1.86	0.35	–35.68	–2.10	0.24	–0.09	0.22	–0.55	–6.77	0.07	–0.23	0.33
Mozambique	12.85	0.97	–66.03	–1.87	0.71	2.44	0.75	–0.08	–8.63	3.80	3.29	0.72
Tanzania	4.20	0.30	–27.16	–2.09	0.25	–0.51	–0.19	–0.62	–9.74	–0.23	–0.60	0.48
Uganda	6.48	1.36	–41.54	–2.38	0.33	0.23	0.20	–0.32	–12.87	–2.14	0.23	0.22
Zambia	3.35	0.35	–38.08	–1.38	0.20	0.13	0.00	–0.48	–9.48	0.25	0.11	0.25
Zimbabwe	1.15	0.20	–22.45	–1.37	0.03	–0.28	0.00	–0.36	–2.97	–0.12	–0.34	0.01
C. With growth in livestock												
Malawi	1.26	0.18	–8.73	–2.10	0.49	1.29	2.74	–0.67	–0.78	0.04	–0.29	0.24
Madagascar	0.51	0.05	–7.20	–1.97	1.05	1.58	4.06	–0.93	–1.32	0.50	1.38	0.67
Mozambique	2.03	0.46	–4.04	–2.23	0.34	0.81	1.56	–0.52	–0.50	0.72	0.72	0.18
Tanzania	1.29	0.38	–4.70	–2.03	0.72	0.52	2.12	–0.78	–1.63	1.03	–0.13	0.98
Uganda	2.02	1.08	–4.49	–2.30	0.81	1.02	1.71	–0.52	–1.49	0.53	0.97	0.23
Zambia	0.75	0.07	–2.69	–1.43	0.27	0.27	2.10	–0.57	–0.94	0.77	–0.11	0.24
Zimbabwe	0.66	0.02	–5.48	–1.24	0.23	1.31	4.57	–0.57	–0.62	–0.02	0.98	–0.01

Source: CGE model results.

[a]For maize and other coarse grains.

Table A12 Results of Simulations 5 and 6: Improving marketing costs and growth in the nonagricultural sector (with fixed labor supply) (additional per capita annual growth rate, %)

	Real GDP	Agri-cultural income	Food consump-tion	Food price index	Agri-cultural imports	Agri-cultural exports	Return to land	Return to unskilled labor	Return to skilled labor
A. Selected agriculture (combination of Scenarios 1, 2, and 4)									
Malawi	0.47	0.89	2.87	–0.78	3.43	1.30	–1.86	0.44	0.27
Madagascar	0.82	1.99	3.95	–0.80	1.13	2.63	1.88	1.16	0.38
Mozambique	0.33	1.39	1.65	–0.45	0.58	4.43	1.43	0.54	0.18
Tanzania	0.76	1.07	2.25	–0.66	1.12	4.52	0.46	0.78	0.92
Uganda	0.66	1.49	1.42	0.18	1.72	4.44	1.62	1.14	0.32
Zambia	0.32	0.68	2.26	–0.60	–0.44	5.88	0.36	0.45	0.34
Zimbabwe	0.27	0.83	4.73	–0.67	1.95	2.25	0.27	0.30	0.12
B. With reducing marketing costs (Scenario 5)									
Malawi	1.13	1.97	3.30	–0.46	8.78	4.17	0.43	0.86	0.48
Madagascar	1.24	2.86	4.14	–0.46	3.36	4.50	2.85	1.64	0.61
Mozambique	0.62	1.99	1.88	–0.19	1.37	5.72	2.08	0.85	0.42
Tanzania	1.13	1.88	2.54	–0.36	2.47	5.67	1.32	1.30	1.06
Uganda	0.95	2.21	1.50	0.61	3.18	6.18	2.42	1.68	0.52
Zambia	1.06	1.90	2.88	–0.35	1.82	8.43	1.69	0.83	0.64
Zimbabwe	0.73	2.11	5.15	–0.45	5.14	4.23	2.06	0.63	0.34
C. With nonagricultural growth (Scenario 6)									
Malawi	2.40	4.38	4.47	0.37	8.50	4.92	3.46	1.99	0.89
Madagascar	1.92	4.05	4.58	0.02	3.93	3.61	3.93	2.32	0.27
Mozambique	1.77	3.87	2.80	0.76	2.22	5.72	3.91	2.19	1.26
Tanzania	1.98	3.45	3.21	0.36	2.44	4.65	2.89	2.47	1.30
Uganda	1.44	3.02	1.69	1.22	3.80	5.11	3.15	2.43	1.21
Zambia	2.12	4.05	3.83	0.37	1.58	7.82	3.96	1.84	0.79
Zimbabwe	1.67	3.92	6.19	0.13	5.50	4.45	4.12	1.55	0.93

Source: CGE model results.

Table A13 Results of Simulations 5 and 6: Improving marketing costs and growth in the nonagricultural sector (with unemployed unskilled labor) (additional per capita annual growth rate, %)

	Real GDP	Agricultural income	Food consumption	Food price index	Agricultural imports	Agricultural exports	Return to land	Return to skilled labor
A. Selected agriculture (combination of Scenarios 1, 2, and 4)								
Malawi	0.70	1.11	3.16	–0.87	3.61	1.58	–1.61	0.47
Madagascar	1.51	2.52	4.74	–1.12	0.19	4.61	2.63	1.12
Mozambique	0.60	1.81	2.04	–0.56	0.38	5.22	2.04	0.37
Tanzania	1.22	1.46	2.84	–0.85	0.85	5.71	1.06	1.43
Uganda	1.44	2.23	2.53	–0.27	1.25	6.57	2.79	0.68
Zambia	0.54	0.98	2.57	–0.69	–0.65	6.93	0.81	0.48
Zimbabwe	0.39	0.87	4.89	–0.72	1.97	2.47	0.37	0.22
B. With reducing marketing costs (Scenario 5)								
Malawi	1.59	2.40	3.88	–0.64	9.07	4.71	0.93	0.86
Madagascar	2.23	3.64	5.26	–0.92	2.03	7.36	3.95	1.66
Mozambique	1.06	2.66	2.51	–0.38	1.07	6.98	3.05	0.72
Tanzania	1.89	2.52	3.53	–0.66	2.01	7.68	2.34	1.91
Uganda	2.11	3.30	3.14	–0.05	2.52	9.37	4.16	1.04
Zambia	1.48	2.49	3.47	–0.53	1.39	10.40	2.56	0.92
Zimbabwe	0.99	2.21	5.48	–0.55	5.15	4.71	2.27	0.56
C. With nonagricultural growth (Scenario 6)								
Malawi	3.47	5.40	5.81	–0.06	9.20	6.18	4.65	1.78
Madagascar	3.32	5.16	6.18	–0.62	2.05	7.65	5.49	1.76
Mozambique	2.91	5.62	4.43	0.28	1.42	8.95	6.45	2.03
Tanzania	3.44	4.70	5.09	–0.22	1.50	8.42	4.86	2.92
Uganda	3.11	4.60	4.07	0.27	2.85	9.68	5.69	1.98
Zambia	3.06	5.39	5.14	0.00	0.46	12.12	5.93	1.41
Zimbabwe	2.31	4.19	7.01	–0.12	5.52	5.64	4.68	1.47

Source: CGE model results.

APPENDIX B

Mathematical Presentation of the CGE Model

Table B1 Country and sector aggregation in the model

Countries and country groups	Commodities and commodity groups
1. Botswana	1. Paddy rice
2. Madagascar	2. Wheat
3. Malawi	3. Other grains (mainly maize)
4. Mozambique	4. Exportable vegetables and fruits
5. Republic of South Africa	5. Domestic vegetables and fruits
6. Tanzania	6. Oilseeds
7. Uganda	7. Sugarcane and sugar beet
8. Zambia	8. Plant-based fibers (mainly cotton)
9. Zimbabwe	9. Other exportable crops (mainly tree crops)
10. Southern African Custom Union	10. Other domestic crops
11. Rest of Southern Africa	11. Cattle, sheep, goats, and horses
12. Rest of Sub-Saharan Africa	12. Animal products
13. Morocco	13. Raw milk
14. Tunisia	14. Forestry
15. Rest of North Africa	15. Fishing
16. China	16. Cattle, sheep, and goat meat products
17. India	17. Other meat products
18. Rest of Asia	18. Vegetable oils and fat
19. United States	19. Dairy products
20. European Union and European Free Trade Area	20. Processed rice
21. Other Europe	21. Sugar
22. Rest of world	22. Processed food products
	23. Beverages and tobacco products
	24. Energy products
	25. Minerals
	26. Textiles and clothing
	27. Heavy industrial products
	28. Machinery and equipment
	29. Electricity, water, and urban utility supply
	30. Wholesale and retail trade services
	31. Transportation
	32. Water and air transportation
	33. Other services
	34. Public administrative, education, and health services

Model Notation

The subscripts i and j in the following variables and parameters refer to commodity/sector set, in which *otp* is a specific element to represent transport sector, f refers to factors, and r and s refer to country/regional group set.

Variables

Production Side

$PX_{i,r}$	Producer price of good i in country r
$PVA_{i,r}$	Value added price of good i in country r
$WF_{f,r}$	Returns to factor f in country r
$X_{i,r}$	Output of sector i produced in country r
$FD_{f,i,r}$	Factor demand of f by sector i in country r
$FS_{f,r}$	Total supply of factor f in country r
$INTD_{i,r}$	Intermediate demand of good i in country r

Demand Side

YH_r	Household total income in country r
$GOVREV_r$	Government revenue in country r
$ZTOT_r$	Total investment in country r
$GOVTRAN_r$	Government transfers to household in country r
$CD_{i,r}$	Household demand of good i in country r
$GD_{i,r}$	Government demand of good i in country r
$INVD_{i,r}$	Investment demand of good i in country r

Trade

$PWM_{i,r,s}$	CIF price of good i for country s imported from r
$PWE_{i,r,s}$	FOB price of good i for country r exporting to country s
$PM_{i,r,s}$	Import price of good i in the domestic market of country s and imported from country r
$PE_{i,r,s}$	Export price of good i at the border of country r and exporting to country s
$PMM_{i,r}$	Armington price of import-composite good i for country r
$PEE_{i,r}$	CET price of export-composite good i in country r
$PD_{i,r}$	Price for output i domestically produced and consumed in country r
$PC_{i,r}$	Armington price of composite good i in country r
$E_{i,r,s}$	Good i exporting from country r to country s
$M_{i,r,s}$	Good i imported by country s from country r
$EE_{i,r}$	Export-composite good i for country r
$MM_{i,r}$	Import-composite good i for country r
$DC_{i,r}$	Output i domestically produced and consumed in country r
$CC_{i,r}$	Composite good i for country r
$TRANSPR_{i,r,s}$	International transport cost for good i shipping from country r to s
$TSPRM_{i,r,s}$	Transport cost for good i imported by country s from country r incurred in the domestic markets of country s
$TSPRE_{i,r,s}$	Transport cost for good i exporting from country r to country s and incurred in the domestic markets of country r

Macro closures

$\overline{FSAVE_r}$	Fixed net foreign savings (trade deficits) of country r
$\overline{GOVEXPS_r}$	Fixed government total expenditure in country r

Parameters

Defined Substitution Elasticities

$\sigma^c_{i,r}$	Armington elasticity of substitution between domestic and import-composite good i in country r
$\sigma^m_{i,r}$	Armington elasticity of substitution between imports of good i in country r imported from different exporting countries
$\sigma^t_{i,r}$	CET elasticity of substitution between domestic and export-composite good i in country r
$\sigma^e_{i,r}$	CET elasticity of substitution between exports of good i in country r exporting to different importing countries
$\sigma^x_{i,r}$	Elasticity of substitution in CES value-added production function for sector i in country r

Computed Substitution Elasticities, Shares, and Shift Parameters

$\beta^c_{i,r}$	Share parameter in household's demand function for good i in country r
$\beta^g_{i,r}$	Share parameter in government's demand function for good i in country r
$\beta^z_{i,r}$	Share parameter in investment demand function for good i in country r
$\alpha^c_{f,i,r}$	Share parameter in value-added production function of sector i for factor f in country r
$\delta^m_{i,r,s}$	Share parameters in Armington import function for good i imported by country s from country r
$\delta^e_{i,r,s}$	Share parameters in CET export function for good i exported by country r to country s
$\delta^t_{i,r}$	Share parameters in CET function for export-composite good i in country r
$\delta_{i,r}$	Share parameters in Armington function for import-composite good i imported in country r
$\gamma_{i,r}$	Subsistence parameter in Stone-Geary utility function
$A^m_{i,r}$	Shift parameter for import good i by country r in Armington import function
$A^c_{i,r}$	Shift parameter for import-composite good i in Armington function in country r
$A^e_{i,r}$	Shift parameter for export good i in country r in CET export function
$A^t_{i,r}$	Shift parameter for export-composite good i in country r in CET function
$A^x_{i,r}$	Shift parameter for sector i in country r in CES value-added production function

Other Computed Parameters

$ø_{i,r,s}$	International transport margin for good *i* exported from country *r* to country *s*
$ø^{m}_{i,r,s}$	Transport margin for imports of good *i* paid to the domestic transport firm of importing country *s* and imported from country *r*
$ø^{e}_{i,r,s}$	Transport margin for exports of good *i* paid to the domestic transport firm of exporting country *r* and imported by country *s*
$io_{i,j,r}$	Input-output coefficient for good *i* used in sector *j* in country *r*
$xtaxr_{i,r,s}$	Export tax rate on good *i* in country *r* exporting to *s*
$mtaxr_{i,r,s}$	Import tax rate on good *i* in country *s* imported from country *r*
$ptaxr_{i,r}$	Producer tax on good *i* in country *r*
$ctaxr_{i,r}$	Commodity sales tax rate on good *i* in country *r*
$hsaver_{r}$	Household saving rates in country *r*
exr_{r}	Nominal exchange rate in country *r*

Figure B1 Illustration of the CGE model: Within-country commodity/factor flow

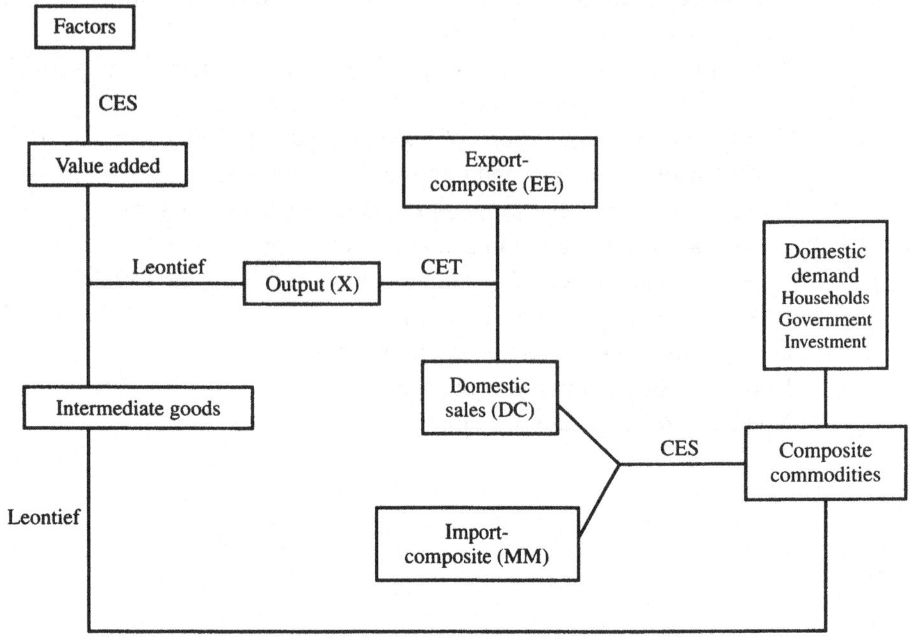

Figure B2 Illustration of the CGE model: Trade flows between countries

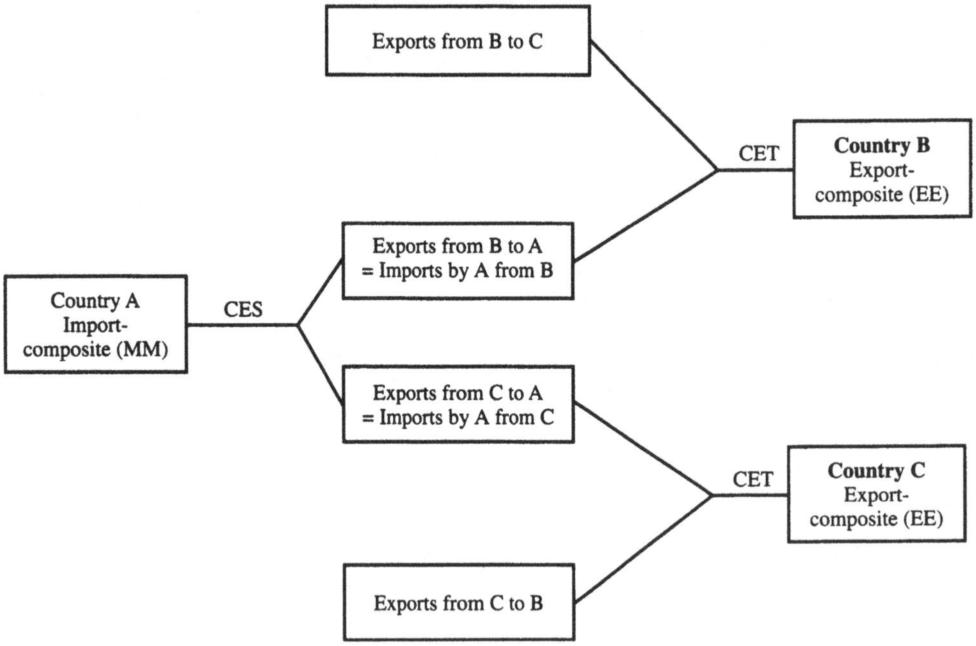

B4. Model Equations

Relationship between CIF and FOB Prices

(1) $PWM_{i,s,r} = (1+TRANSPR_{i,s,r}) \times PWE_{i,s,r}$

(1a) $TRANSPR_{i,s,r} = \emptyset_{i,s,r} \times PWE_{otp,s,r}$

Production and Input Demand

CES value-added function

$$(2) \quad PVA_{i,s,r} = \left(\Lambda^{x}_{i,r}\right)^{-1} \times \left[\sum_{f} \left(\alpha^{\sigma^{x}_{i,r}}_{f,i,r} \times WF^{1-\sigma^{x}_{i,r}}_{f,r}\right)\right]^{\frac{1}{(1-\sigma^{x}_{i,r})}}$$

Factor demand

$$(3) \quad FD_{f,i,r} = \left(\Lambda^{d}_{i,r}\right)^{\sigma^{x}_{i,r}-1} \times \left(\frac{PVA_{i,r} \times \alpha_{f,i,r}}{WF_{f,r}}\right)^{\sigma^{x}_{i,r}} \times X_{i,r}$$

Intermediate demand

$$(4) \quad INTD_{i,r} = \sum_{i'}\left(io_{i,i',r} \times X_{i,r}\right)$$

Relationship between value-added and output prices

$$(5) \quad \frac{PX_{i,r}}{(1 + ptaxr_{i,r})} = PVA_{ir} + \sum_{i'}\left[io_{i,i',r} \times PC_{i',r} \times (1 + ctaxr_{i',r})\right]$$

Imports and Exports

Armington import function for composite goods

$$(6) \quad PC_{i,r} = \left(\Lambda^{c}_{i,r}\right)^{-1} \times \left(\delta^{\sigma^{c}_{i,r}}_{i,r} \times PMM^{1-\sigma^{c}_{i,r}}_{i,r} + (1 - \delta_{i,r})^{\sigma^{c}_{i,r}} PD^{1-\sigma^{c}_{i,r}}_{i,r}\right)^{\frac{1}{(1-\sigma^{c}_{umr})}}$$

Demand for import-composite goods

$$(7) \quad MM_{i,r} = \left(\Lambda^{c}_{i,r}\right)^{\sigma^{c}_{i,r}-1} \times \left(\frac{\delta_{i,r} \times PC_{i,r}}{PMM_{i,r}}\right)^{\sigma^{c}_{i,r}} \times CC_{i,r}$$

Demand for domestically produced goods

$$(8) \quad DC_{i,r} = \left(\Lambda^{c}_{i,r}\right)^{\sigma^{c}_{i,r}-1} \times \left(\frac{(1 - \delta_{i,r}) \times PC_{i,r}}{PD_{i,r}}\right)^{\sigma^{c}_{i,r}} \times CC_{i,r}$$

Armington function for import-composite goods

$$(9) \quad PMM_{i,r} = \left(\Lambda^{m}_{i,r}\right)^{-1} \times \left[\sum_{s}\left(\left(\delta^{m}_{i,s,r}\right)^{\sigma^{m}_{i,r}} \times PM^{1-\sigma^{m}_{i,r}}_{i,s,r}\right)\right]^{\frac{1}{(1-\sigma^{m}_{i,r})}}$$

Import price in domestic markets

$$(10) \quad PM_{i,r} = (1 + mtaxr_{i,s,r}) \times EXR_{r} \times (1 + TSPRM_{i,s,r,s}) \times PWM_{i,s,r}$$

$$(10a) \quad TSPRM_{i,s,r} = \phi_{i,s,r} \times PX_{otp',r}$$

Imports demand

$$(11)\quad M_{i,s,r} = \left(\Lambda^{m}_{i,r}\right)^{\sigma^{m}_{i,s} - 1} \times \left(\frac{PMM_{i,r} \times \delta^{m}_{f,i,r}}{PM_{i,s,r}}\right)^{\sigma^{m}_{i,s}} \times MM_{i,r}$$

CET function for export-composite goods

$$(12)\quad PX_{i,r} = \left(\Lambda^{t}_{i,r}\right)^{-1} \times \left(\left(\delta^{t}_{i,r}\right)^{-\sigma^{t}_{i,r}} \times PEE_{i,r}^{1+\sigma^{t}_{i,r}} + \left(1 - \delta^{t}_{i,r}\right)^{-\sigma^{t}_{i,r}} PD_{i,r}^{1+\sigma^{t}_{i,r}}\right)^{\frac{1}{(1+\sigma^{t}_{i,r})}}$$

Supply of export-composite goods

$$(13)\quad EE_{i,r} = \left(\Lambda^{t}_{i,r}\right)^{-\left(1+\sigma^{t}_{i,r}\right)} \times \left(\frac{\left(1 - \delta^{t}_{i,r}\right) \times PX_{i,r}}{PD_{i,r}}\right)^{-\sigma^{t}_{i,r}} \times X_{i,}$$

Supply to domestic markets

$$(14)\quad DC_{i,r} = \left(\Lambda^{t}_{i,r}\right) -^{\left(1+\sigma^{t}_{i,r}\right)} \times \left(\frac{\left(1 - \delta^{t}_{i,r}\right) \times PX_{i,r}}{PD_{i,r}}\right)^{-\sigma^{t}_{i,r}} \times X_{i,r}$$

CET function of export-composite goods

$$(15)\quad PEE_{i,r} = \Lambda^{e}_{i,r} \times \left(\left(\delta^{e}_{i,r}\right)^{-\sigma^{e}_{i,r}} \times PE_{i,r,s}^{1+\sigma^{e}_{i,r}}\right)^{\frac{1}{1+\sigma^{e}_{i,r}}}$$

Export price in domestic markets

$$(16)\quad PE_{i,r,s} = \frac{\left(1 - xtaxr_{i,r,s}\right) \times EXR_{r}}{\left(1 + TSPRE_{i,r,s}\right)} \times PWE_{i,r,s}$$

$$(16a)\quad TSPRE_{i,r,s} = \phi^{e}_{i,r,s} \times PX_{'otp',r}$$

Export supply

$$(17)\quad E_{i,r,s} = \left(\Lambda^{e}_{i,r}\right)^{-\left(1+\sigma^{e}_{i,r}\right)} \times \left(\frac{PEE_{i,r} \times \delta^{e}_{i,r,s}}{PE_{i,r,s}}\right)^{-\sigma^{e}_{i,r}} \times EE_{i,r}$$

Identification between imports by country *r* from country *s* and exports from country *s* to country *r*

$$(18)\quad M_{i,s,r} = E_{i,s,r}$$

Final Demand and Income

Household income

$$(19)\quad YH_{r} = \sum_{f}\sum_{i} WF_{f,r} \times FD_{f,i,r} \times GOVTRAN_{r}$$

Household consumption demand

$$(20)\quad CD_{i,r} = \frac{\beta^{c}_{i,r} \times \left(YH_{r} \times \left(1 - hsaver_{r}\right) - \sum_{i'} PCT_{i',r \times \gamma_{i,r}}\right)}{PC_{i,r} \times \left(1 + ctaxr_{i,r}\right)} + \gamma_{i,r}$$

Government revenue

$$(21)\ GOVREV_r = \sum_i \sum_s [xtaxr_{i,r,s} \times exr_r \times PWE_{i,r,s} \times E_{i,r,s}] + \sum_i \sum_s [mtaxr_{i,r,s} \times exr_r \times PWM_{i,s,r} \times M_{i,s,r}] + \sum_i [PC_{i,r} \times (1 + ctaxr_{i,r}) \times CC_{i,r}] + \sum_i \{[ptaxr_{i,r} / (1 + ptaxr_{i,r})] \times PX_{i,r} \times X_{i,r}\}$$

Government final demand

$$(22)\ GD_{i,r} = \frac{\beta^{g}_{i,r} \times \overline{GOVEXPS_r}}{(PC_{i,r} \times (1 + ctaxr_{i,r}))}$$

Government transfers

$$(23)\ GOVTRAN_r = GOVREV_r - \overline{GOVEXPS_r}$$

Investment demand

$$(24)\ INVD_{i,r} = \frac{\beta^{z}_{i,r} \times ZTOT_r}{(PC_{i,r} \times (1 + ctaxr_{i,r}))}$$

Equilibrium conditions

Commodity markets

$$(25)\ CC_{ir} = CD_{i,r} + GD_{i,r} + INVD_{i,r} + INTD_{ir}$$

Factor market

$$(26)\ \sum_i FD_{f,i,r} = FS_{f,r}$$

Foreign savings

$$(27)\ \overline{FSAVE_r} = \sum_i \sum_s (PWM_{i,s,r} \times M_{i,s,r}) - \sum_i \sum_s (PWE_{i,r,s} \times E_{i,r,s})$$

References

Adelman, I. 1984. Beyond export-led growth. *World Development* 12 (9): 937–949.

AFDB (African Development Bank) and OECD (Organization for Economic Cooperation and Development). 2002. *Achieving the millennium development goals in Africa: Progress, prospects, and policy implications*. Paris: OECD.

Akiyama, T., J. Baffes, D. F. Larson, and P. Varangis. 2003. *Commodity market reform in Africa*. Policy Research Working Paper 2995. Washington, D.C.: World Bank.

Ashley, C., and S. Maxwell. 2001. Rethinking rural development. *Development Policy Review* 19 (4): 395–425.

Bacon, C. (2005). Confronting the coffee crisis: Can fair trade, organic, and specialty coffees reduce small-scale farmer vulnerability in northern Nicaragua? *World Development* 33 (3): 497–511.

Baffes, J. 2003. *Cotton, market setting, policies, issues, and facts*. Policy Research Working Paper. Washington, D.C.: World Bank.

———. 2006. Restructuring Uganda's coffee industry: Why going back to basics matters. *Development Policy Review* 24 (4): 413–436.

Baffes, J., and B. Gardner. 2003. The transmission of world commodity prices to domestic markets under policy reforms in developing countries. *Journal of Policy Reforms* 6 (3): 159–180.

Balassa, B. 1965. Trade liberalization and "revealed" comparative advantage. *The Manchester School of Economic and Social Studies* 33: 99–123.

Barrett, H., A. Brown, B. Ilbery, G. Jackson, and T. Bins. 1997. *Prospect for horticultural exports under trade liberalization in adjusting African economies*. Report submitted to the Department for International Development, Coventry University, Coventry, U.K.

Bell, C., and P. B. R. Hazell. 1980. Measuring the indirect effects of an agricultural investment project on its surrounding region. *American Journal of Agricultural Economics* 62: 75–86.

Binns, T., and N. Fereday. 1996. Feeding African urban poor: Urban and peri-urban horticulture in Kano, Nigeria. *Geography* 81(4): 380–384.

Burfisher, M. E., S. Robinson, and K. Thierfelder. 2002. The global impacts of farm policy reforms in Organization for Economic Cooperation in developing countries. *American Journal of Agricultural Economics* 84 (3): 234–241.

Byerlee, D., and C. Eicher. 1997. *Africa's emerging maize revolution*. Boulder, Colo., U.S.A.: Lynne Rienner Publishers.

Calo, M., and T. A. Wise. 2005. *Revaluing peasant coffee production: Organic and free trade markets in Mexico*. Medford, Mass., U.S.A.: Global Development and Environment Institute, Tufts University.

Cardenas, M. 1994. Stabilization and redistribution of coffee revenues: A political economy model of commodity marketing boards. *Journal of Development Economics* 44: 351–380.

Collier, P. 2005. Is agriculture still relevant to poverty reduction in Africa? Speech given to All Party Parliamentary Group on Overseas Development, October 17, 2005, House of Commons, London, U.K. <http://www.odi.org.uk/speeches/apgood_oct05/apgood_oct17/report.html>. Accessed August 2006.

Collins, J. 1995. Gender and cheap labor in agriculture. In *Food and agrarian orders in the world economy*, ed. P. McMichael. Westport, Conn., U.S.A.: Praeger.

DECPG (Development Economics Prospects Group, of the World Bank). 2006. Commodity Markets Review, July 12, 2006. < http://decpg.worldbank.org>. Accessed August 2006.

Dehn, J. 2000. *Commodity price uncertainty in developing countries*. Policy Research Working Paper 2426. Washington, D.C.: World Bank.

Dervis, K., J. de Melo, and S. Robinson. 1982. *General equilibrium models for development policy*. World Research Publication. Washington, D.C.: World Bank.

Diao, X., and Y. Yanoma. 2003. *Exploring regional dynamics in Sub-Saharan African agriculture*. Development Strategy and Governance Division (DSGD) Discussion Paper 2. Washington, D.C.: International Food Policy Research Institute.

Diao, X., T. Roe, and E. Yeldan. 1999. Strategic policies and growth: An applied model of R&D-driven endogenous growth. *Journal of Development Economics* 60: 343–380.

Diao, X., T. Roe, and A. Somwaru. 2002. Developing country interests in agricultural reforms under the World Trade Organization. *American Journal of Agricultural Economics* 84 (3): 782–790.

Diao, X., P. Hazell, D. Resnick, and J. Thurlow. 2006. *The role of agriculture in development: Implications for Sub-Saharan Africa*. Development Strategy and Governance Division Discussion Paper 29. Washington, D.C.: International Food Policy Research Institute.

Dolan, C., and J. Humphrey. 2000. *Governance and trade in fresh vegetables: The impact of U.K. supermarkets on the African horticulture industry*. Proceedings, 2000 Conference. University of Sussex, U.K.: Institute of Development Studies.

Dolan, C., J. Humphrey, and C. Harris-Pascal. 1999. *Horticultural commodity chains: The impact of the U.K. market on the African fresh vegetable industry*. Working Paper 96. University of Sussex, U.K.: Institute of Development Studies.

Dorosh, P. A. 2001. Trade liberalization and national food security: Rice trade between Bangladesh and India. *World Development* 29 (4): 673–690.

Dorosh, P., and S. Haggblade. 2003. Growth linkages, price effects and income distribution in Sub-Saharan Africa. *Journal of African Economies* 12 (2): 207–235.

Dorosh, P., D. E. Sahn, and S. D. Younger. 1996. Exchange rate, fiscal and agricultural policies in Africa: Does adjustment hurt the poor? *World Development* 24 (4): 719–747.

Dorward, A., J. Kydd, and C. Poulton. 2005. Beyond liberalization: Developmental coordination policies for African smallholder agriculture. *IDS Bulletin* 36 (2): 80–85.

Dréze, J., and A. Sen. 1989. *Political and public action*. Oxford: Clarendon Press.

Ellis, F. 2005. "Small farms, livelihood diversification, and rural-urban transitions: strategic issues in Sub-Saharan Africa." In *The future of small farms*. Proceedings of a workshop held by the International Food Policy Research Institute/2020 Vision Initiative, Overseas Development Initiative, and Imperial College London, June 26-29, 2005, in Wye, United Kingdom. Washington, D.C.: International Food Policy Research Institute.

FAO (Food and Agriculture Organization of the United Nations). 2000. Export crop liberalization in Africa. *FAO Agricultural Services Bulletin* 135.

———. 2002. World agriculture: towards 2015/2030. Summary Report. Rome: FAO.

FAOSTAT. 2006. FAO Statistical Database. Food and Agricultural Organization of the United Nations, Rome, Italy. <http://faostat.fao.org/>. Accessed March 2006.

Fitter, R., and R. Kaplinsky. 2001. *Who gains from product rents as the coffee market becomes more differentiated? A value chain analysis.* IDS Bulletin Paper. University of Sussex, Brighton, U.K.

Gabre-Madhin, E. Z. 2001. *Market institutions, transaction costs, and social capital in the Ethiopian grain market.* Research Report 124. Washington, D.C.: International Food Policy Research Institute.

GTAP (Global Trade Analysis Project). 2005. *GTAP database, v6* (in CD). West Lafayette, Ind., U.S.A.: Purdue University.

Haggblade, S., J. Hammer, and P. B. R. Hazell. 1991. Modeling agricultural growth multipliers. *American Journal of Agricultural Economics* 73(2): 361–374.

Hallam, D., P. Liu, G. Lavers, P. Pilkauskas, G. Rapsomanikis, and G. Claro. 2004. *The market for non-traditional agricultural exports.* FAO Commodities and Trade Technical Paper 3. Rome: FAO.

Harris, C., P.V. Hegarty, M. Kherallah, C. Mukindia, J. Ngige, P. Sterns, and J. Tatter. 2001. The impacts of standards on the food sector of Kenya. Report prepared by the Institute for Food and Agricultural Standards, Michigan State University, for the U.S. Agency for International Development. East Lansing: Michigan State University.

Harris, S. 1992. *Kenya horticulture sub-sector survey.* Nairobi, Kenya: Export Development Support Project..

Hazell, P. 2005. *Does policy research matter? A farewell lecture*, November 21, 2005. Washington, D.C.: International Food Policy Research Institute.

Henson, S., A. Brouder, and W. Mitullah. 2000. Food safety requirements and food exports from developing countries: The case of fish exports from Kenya to the European Union. *American Journal of Agricultural Economics* 82 (5): 1159–1169.

Humphrey, J., N. McCulloch, and M. Ota. 2004. The impact of European market changes on employment in the Kenyan horticulture sector. *Journal of International Development* 16(1): 63–80.

ICAC (International Cotton Advisory Committee). 2005. <http://www.icac.org>. Accessed June 2005.

ICCO (International Cocoa Organization). 2006. <http://www.icco.org>. Accessed July 14, 2006.

ICO (International Coffee Organization). 2006. <http://www.ico.org>. Accessed July 14, 2006.

ICRISAT. 2001. Sparking a grey-to-green revolution for dry areas: 28 years of ICRISAT's partnership-based research across the semi-arid tropics. <http://www.icrisat.org/text/news/2001/grey-to-green.htm>. Accessed June 2005.

IMF/WB Development Committee. 2001. *Leveraging Trade for Development: World Bank Role.* Mimeo: p. 3.

Islam, N. 1990. *Horticultural exports of developing countries: Past performances, future prospects, and policy issues.* Research Report 80. Washington, D.C.: International Food Policy Research Institute.

Jayne, T. S., M. Mukumbu, M. Chisvo, D. Tschirley, M. T. Weber, B. Zulu, R. Johansson, P. Santos, and D. Soroko. 1999. Successes and challenges of food market reform: Experiences from Kenya, Mozambique, Zambia, and Zimbabwe. MSU International Development Working Paper No. 72. East Lansing: Michigan State University.

Karanja, A. 2002. *Liberalisation and smallholder agriculture development: A case study of coffee farms in Central Kenya.* Working Paper. Wageningen, Netherlands: Wageningen University.

Kherallah, M., C. Delgado, E. Gabre-Madhin, N. Minot, and M. Johnson. 2002. *Reforming agricultural markets in Africa.* Baltimore: Johns Hopkins University Press.

Krivonos, E. 2004. *The impact of coffee market reforms on producer prices and price transmission.* Policy Research Working Paper 3358. Washington, D.C.: World Bank.

Limao, N., and A. J. Venables. 2001. Infrastructure, geographical disadvantage, transport cost, and trade. *World Bank Economic Review* 15(3): 451–479.

Little, P. D., and C. S. Dolan. 1998. What it means to be restructured: Non-traditional commodities and structural adjustment in Sub-Saharan Africa. In *Rethinking commodities: Anthropological views of the global market place*, eds. P. Stone, A. Haugerud, and P. D. Little. Madison, Wis., U.S.A.: University of Wisconsin Press.

Maxwell, S., and R. Slater. 2003. Food policy old and new. *Development Policy Review* 21 (5–6): 531–553.

McCulloch, N., and M. Ota. 2002. *Export horticulture and poverty in Kenya*. Working paper. Brighton, U.K.: University of Sussex, Institute of Development Studies.

Minot, N., and M. Ngigi. 2002. *Are horticultural exports a replicable success story? Evidence from Kenya and Côte d'Ivoire*. Working Paper. Washington, D.C.: International Food Policy Research Institute.

———. 2003. Are horticultural exports a replicable success story? Evidence from Kenya and Côte d'Ivoire. Paper presented at the International Conference of Successes in African Agriculture: Building for the Future, December 1–3, 2003, Pretoria, South Africa.

Mundlak, Y., and D. F. Larson. 1992. On the transmission of world agricultural prices. *The World Bank Economic Review* 6 (3): 399–422.

Muradian, R., and W. Pelupessy. 2005. Governing the coffee chain: The role of voluntary regulatory systems. *World Development* 33 (12): 2029–2044.

Ng, F., and A. Yeats. 2002. *What can Africa expect from its traditional exports?* Africa Region Working Paper 26. Washington, D.C.: World Bank.

Nin Pratt, A., and X. Diao. 2006. *Exploring growth linkages and market opportunities for agriculture in Southern Africa*. Development Strategy and Governance Division Discussion Paper 42. Washington, D.C.: International Food Policy Research Institute.

Noman, A., and J. Stiglitz. 2006. IPD's Africa Task Force: Issues note. A note for the meeting to launch the Task Force on August 3–4, 2006, at Brooks World Poverty Institute, University of Manchester, U.K.

Oyejide, T. A., E. O. Ogunkola, and S. A. Bankole. 2000. *Quantifying the trade impact of sanitary and phytosanitary standards: What is known and issues of importance for Sub-Saharan Africa*. Paper prepared for the workshop on Quantifying the Trade Effect of Standards and Regulatory Barriers: Is It Possible? Washington, D.C.: World Bank.

Ponte, S. 2001. *The "Latte Revolution"? Winners and losers in the restructuring of the global coffee marketing chain*. Working Paper. Copenhagen: Center for Development Research.

Poulton, C., J. Kydd, and A. Dorward. 2006. Overcoming market constraints on pro-poor agricultural growth in Sub-Saharan Africa. *Development Policy Review* 24 (3): 243–277.

Quiroz, J., and R. Soto. 1995. *International price signals in agricultural prices: Do governments care?* Mimeo. Washington, D.C.: World Bank.

Reardon, T., C. P. Timmer, C. Barrett, and J. Berdegue. 2003. The rise of supermarkets in Africa, Asia, and Latin America. *American Journal of Agricultural Economics* 85 (5): 1140–1146.

Reimer, J. J., and T. W. Hertel. 2004. *International cross section estimates of demand for use in the GTAP model*. GTAP Technical Paper 23. West Lafayette, Ind., U.S.A.: Purdue University.

Rosegrant, M. W., M. Paisner, S. Meijer, and J. Witcover. 2001. *Global food projects to 2020*. Washington, D.C.: International Food Policy Research Institute.

Sanders, J., B. Shapiro, and S. Ramasamy. 1996. *The economics of agricultural technology in semiarid Sub-Saharan Africa*. Baltimore: Johns Hopkins University Press.

SIDA (2006). *What can Sub-Saharan Africa learn from Asian experiences in addressing its food crisis?* Stockholm: Swedish International Development Cooperation Agency, Division for Rural Development.

Singh, B. P. 2002. Nontraditional crop production in Africa for export. In *Trends in new crops and new uses*, eds. J. Janick and A. Whipkey. Alexandria, Va., U.S.A.: ASHS Press.

Skees, J., P. Hazell, and M. Miranda. 1999. *New approaches to crop yield insurance in developing countries.* Environment and Production Technology Division Discussion Paper 55. Washington, D.C.: International Food Policy Research Institute.

Stevens, C. 2003. Food trade and food policy in Sub-Saharan Africa: Old myths and new challenges. *Development Policy Review* 21 (5–6): 669–681.

Talbot, J. M. 1995–96. Regulating the coffee commodity chain: Internationalization and the coffee cartel. *Berkley Journal of Sociology* 40: 113–149.

———. 1997. Where does your coffee dollar go?: The division of income and surplus along the coffee commodity chain. *Studies in Comparative International Development* 32 (1): 56–91.

Temu, A., and A. Winter-Nelson. 2002. *Institutional adjustment and agricultural markets: Following the transaction costs in the Tanzanian coffee system. World Development* 30(4): 561–574.

Thoen, R., S. Jaffee, C. Dolan, and F. Ba. 2001. Equatorial rose: *The Kenyan–European cut flower supply chain.* Working Paper. Washington, D.C.: World Bank.

Townsend, T., and A. Guichounts. 1994. *A survey of cotton income and price support programs.* Washington, D.C.: Proceedings of Belt-wide Cotton Conferences.

UNCOMTRADE (United Nations Commodity Trade Statistics Database). 2002. Rev 3.

———. 2005. http://unstats.un.org/unsd/comtrade. Accessed April 2005.

UNCTAD (United Nations Conference on Trade and Development). 2001. *Economic development in Africa: Performance, prospects, and policy issues.* New York, Geneva.

Varangis, P., and G. Schreiber. 2001. Coffee market reforms in West Africa. In *Commodity market reforms: Lessons of two decades*, eds. T. Akiyama, J. Baffes, D. F. Larson, and P. Varangis. Washington, D.C.: World Bank.

Von Braun, J. 1994. Nontraditional vegetable crops and food security among smallholder farmers in Guatemala. In *Agricultural commercialization, economic development, and nutrition.* Baltimore: Johns Hopkins University Press.

WARDA. 2006. NERICA on the Move. <http://www.warda.org/warda/uplandnerica.asp>. Accessed June 2005.

Weatherspoon, D. D., and T. Reardon. 2003. The rise of supermarkets in Africa: Implications for agrifood systems and the rural poor. *Development Policy Review* 21 (3): 333–355.

World Bank. 2004. *Global poverty down by half since 1981, but progress uneven as economic growth eludes many countries.* News Release No. 2004/309/S,23 April 2004. <http://www.worldbank.org/data/wdi2004/index.htm>. Accessed June 2005.

World Bank. 2006. World Development Indicators. Washington, D.C.: World Bank. CD-ROM.

Yeats, A. J. 1998. What can be expected from African regional trade arrangements? Some empirical evidence. Unpublished memorandum. Washington, D.C.: World Bank.

Made in the USA
Monee, IL
08 November 2026

67984624R00057